The Truth About Grief

*The Myth of Its Five Stages
and the New Science of Loss*

Ruth Davis Konigsberg

Simon & Schuster
New York London Toronto Sydney

Simon & Schuster
1230 Avenue of the Americas
New York, NY 10020

First Simon & Schuster hardcover edition January 2011

SIMON & SCHUSTER and colophon are registered trademarks
of Simon & Schuster, Inc.

For information about special discounts for bulk purchases,
please contact Simon & Schuster Special Sales at
1-866-506-1949 or business@simonandschuster.com.

The Simon & Schuster Speakers Bureau can bring authors
to your live event. For more information or to book an event,
contact the Simon & Schuster Speakers Bureau at
1-866-248-3049 or visit our website at www.simonspeakers.com.

Designed by Jill Putorti

Manufactured in the United States of America

10   9   8   7   6   5   4   3   2   1

Library of Congress Cataloging-in-Publication Data is available.

ISBN 978-1-4391-4833-4
ISBN 978-1-4391-5264-5 (ebook)

Graphs on pages 71 and 72 are from "Social support as a predictor of variability:
An examination of the adjustment trajectories of recent widows," by Toni L.
Bisconti, Cindy S. Bergman, and Steven M. Boker, in *Psychology and Aging*,
vol. 21(3), Sept. 2006, 590–99, doi: 10.1037/0882-7974.21.3.590. Copyright © 2006
by the American Psychological Association. Reproduced with permission.
The use of APA information does not imply endorsement by APA.

For my family

Contents

The Truth
About Grief

Introduction

~~~~~

The Idea That Won't Die

In 2008, after Barack Obama opened an irretrievable lead over Hillary Clinton in the Democratic presidential primaries, Lanny Davis, a die-hard Clinton supporter who had also served as special counsel in her husband's administration, said that he was so distraught that he had Googled Elisabeth Kübler-Ross's five stages of grief. "Denial, yes," he said. "Anger, definitely. Bargaining, well, O.K. And depression, that's definitely what I was going through." It wasn't until Obama's acknowledgment of Hillary Clinton in his speech at the convention that Davis said he reached the last stage, acceptance.

In 2009, after an investigation found that a large debt collection agency had been pursuing outstanding payments of people who had recently died by calling up their next of kin, the company's CEO defended the practice in the press. After all, his team of three hundred collectors, he said, were "all trained in the five stages of grief."

In 2010, after NBC dumped Conan O'Brien as the host of *The Tonight Show* and reinstalled Jay Leno, Conan joked on his subsequent comedy tour that visits to a psychiatrist helped him to see that there were stages to the loss of a talk show not unlike the stages of grieving. That same year, commentators invoked the stages to describe our emotional reactions to everything from the TV show *Lost* going off the air to the damage to the Gulf of Mexico after the BP oil spill.

Once you start looking, the stages seem to pop up everywhere. They've become a stock reference in popular entertainment, turning up in episodes of *Frasier* and *The Simpsons,* and more recently *The Office, Grey's Anatomy, Scrubs,* and *House.* They're continuously employed as a literary device—Frank Rich has used them in his *New York Times* opinion column no fewer than five times, such as his remark in 2008 about the occupation of Iraq that "this war has lasted so long that Americans . . . have had the time to pass through all five of the Kübler-Ross stages of grief over its implosion." There's even an acronym to help you remember their sequence: Dabda. In 2008, Sotheby's auctioned a large painting by the British artist Damien Hirst titled *D,A,B,D,A* that consisted of five different colored panels overlaid with real butterflies. (Kübler-Ross loved butterflies and often likened death to a butterfly shedding its cocoon.) The painting sold for $2,650,818.

The stages are so pervasive that they have become axiomatic, divorced from the specific time and place of their origin, but they made their debut in 1969 with the publication of Elisabeth Kübler-Ross's first book, *On Death and Dying*, in which she argued that all people grapple with the end of life by traversing denial, anger, bargaining, depression, and acceptance. If you ignored or repressed the stages, you risked getting stuck with unresolved and painful emotions. But if you plunged yourself through them, you would eventually emerge on the other side stronger and wiser, a reward that was particularly appealing in the 1970s as the self-help movement with its promises of personal transformation was sweeping the country. The book was a surprise bestseller, and Kübler-Ross, then a staff psychiatrist at Billings Hospital in Chicago, became an overnight sensation, attracting hundreds to her speaking engagements. Her theory was soon taught in medical and nursing schools and undergraduate classes, and helped launch the new academic discipline of death education.

Kübler-Ross was heralded as a revolutionary who shattered the stoic silence that had surrounded death since World War I, and her efforts certainly lowered barriers and raised the standard of care for dying people and their families. But she also ushered in a distinctly secular and psychological approach to death, one in which the focus shifted from the salvation of the deceased's soul (or at least its transition to some kind of afterlife) to the quality of his or her last days along with the well-being of the survivors.

The hospice movement was already under way, pioneered by a British doctor named Cecily Saunders, who founded St. Christopher's, the first modern center devoted to the dying, in London in 1967. Florence Wald, the dean of Yale Nursing School, spent a year at St. Christopher's and subsequently opened the first hospice in the United States in New Haven in 1971. These two women championed the need for a humane setting in which the terminally ill could prepare themselves for death, and their contributions undeniably changed end-of-life care for the better. But it wasn't long before a solution was put forth to help bereaved families as well, one promoted by an entirely new professional group specializing in the task of mitigating grief's impact. As I explain in Chapter 5, counseling for grief, though well-intentioned, does not, on average, seem to hasten its departure, and some even think it can harm instead of heal. (This doesn't mean that no one is ever helped by counseling, but that it doesn't measurably benefit its recipients overall when compared to groups that don't receive formalized help.) In retrospect, the practice suffered from becoming popular before there was enough solid research on normal grief to base it upon (most of the existing literature consisted of extreme case studies drawn from clinic populations). From the 1970s to the 1990s, thousands entered the field, setting up healing centers and offering individual counseling or hosting support groups at hospitals, churches, and even funeral homes. These counselors introduced their own theories, turning anecdotal descriptions into treatment plans and modifying Kübler-Ross's stages into a series of phases, tasks, or needs that required active participation, as well as outside professional help. In this increasingly complex emotional landscape, grief became a "process," or a "journey" to be completed, as well as an opportunity for growth. Few questioned the necessity of a large corps of private counselors dedicated to grief, despite the fact that no country other than the United States seemed to have one. Our modern, atomized society had been stripped of religious faith and ritual and no longer provided adequate support for the bereaved. And so a new belief system rooted in the principles of psychotherapy rose up to help organize the experience. As this system grew more firmly established, it also became more orthodox, allowing for less variation in how to approach the pain and sorrow of loss. By the end of the 1990s, it had become conventional wisdom that people had to explore and give voice to their grief or else it would fester.

Paradoxically, this close examination and enumeration of grief did not bring much greater clarity to specific characteristics of the experience. In 1984, an Institute of Medicine report concluded that a lack of a reliable way to measure grief was a major barrier to being able to help the bereaved. Since then, practitioners have struggled to catalogue all the manifestations of the emotional and psychological upheaval that occurs after a loved one dies. There are now more than twenty different "instruments" (questionnaires) out there—from the Texas Revised Inventory of Grief to the Hogan Grief Reaction Checklist—with anywhere from six to sixty-seven different "items" (symptoms) on them, such as "I have little control over my sadness" or "I frequently feel bitter" or "I am stronger because of the grief I have experienced." This lack of an agreed-upon definition for grief did not slow down the stream of theories on how to best manage the suffering it caused.

Such was our environment when, on September 11, 2001, terrorist attacks killing almost three thousand people irrevocably transformed grief from a private experience into a public, communal one. The loss of those lives was collective: we were all attacked, and we all mourned, attending candlelight vigils, leaving flowers and other tokens of sympathy at spontaneous memorial sites, displaying bumper stickers and T-shirts and baseball caps with the refrain, "Never Forget." The government harnessed this mass mourning to gather support for invading Afghanistan and later Iraq, but soon grief became the source for antiwar protest, as mothers of slain soldiers such as Cindy Sheehan demanded that the president justify her sorrow, challenging the wartime ideology that her son's death, in the service of his country, was for the greater good.

This new emphasis on the public expression of individual loss came to dominate civilian grief as well. Sociologists (and state highway officials) noted an increase in makeshift shrines such as those found at car crash sites on the sides of roads, or spray-painted on the walls of inner city streets, inviting an audience from any and all passersby. Web memorials and online obituaries where people without any relationship to the deceased could post their condolences further blurred the public and private domains of grief. (A 2002 survey of the guest book entries at Worldwidecemetery.com found that 42 percent had been written by strangers.) First-person accounts of widowhood such as Joan Didion's *The Year of Magical Thinking* and Kate Braestrup's *Here If You Need Me* had lengthy stays on bestseller lists. A number of novels and TV shows with a focus on death and the afterlife became popular: *The Lovely Bones, The Five People You Meet in Heaven, The Shack, Six Feet Under, CSI, Rescue Me.*

Traffic in personal grief narratives became increasingly congested as ordinary citizens, given a voice by new media, began disclosing their own experiences on blogs, podcasts, and Internet radio shows. These accounts, while genuine and moving, were also based on the assumed therapeutic value of such public airing. "Telling your story often and in detail is primal to the grieving process," Kübler-Ross had advised. "You must get it out. Grief must be witnessed to be healed."

Entrepreneurs seized on the commercial possibilities of this mandate and opened up grief retreats, where you can get grief massages or do grief yoga. And the self-improvement shelves of the bookstore grew heavier not just with advice on how to survive loss but also grief workbooks and journals, illustrating just how prescribed our emotional behavior after the death of a loved one had become. As Tony Walter, a British sociologist, has written, "Contemporary bereavement is a matter of self-monitoring, assisted by advice from family and friends, bereavement books, counselors and mutual help groups. In this, bereavement is like contemporary marriage and child-rearing in which partners and parents are always asking how well they are doing, consulting the baby books to see if their child's development is above or below average." We never seemed to notice how grief had been shaped by all these social and cultural forces, in part because we had been told that our way of grieving was natural and instinctual, and therefore the best way.

The first I heard of Kübler-Ross's five stages was in 1985, in a high school psychology class, although we were not actually assigned to read *On Death and Dying*. (The teacher was moonlighting from his usual role as wrestling coach.) If I had read the book then, I would have learned that Kübler-Ross was actually writing about the experience of facing one's *own* death, not the death of someone else. It was other practitioners, having found the stages so irresistibly prescriptive, who began applying them to grief in the 1970s, a repurposing that Kübler-Ross did not object to. "Any natural, normal human being, when faced with any kind of loss, will go from shock all the way through acceptance," she said in an interview published in 1981. "You could say the same about divorce, losing a job, a maid, a parakeet." Decades passed before Kübler-Ross decided that it was finally time to properly claim the stages of grief as well. Her nineteenth and final book, *On Grief and Grieving,* was published in 2005, a year after her own death.

"One of the reasons for writing *On Grief and Grieving* was that everyone else had already adapted the stages of dying to the stages of grief," her co-author, David Kessler, told me when I contacted him in 2007. "She always knew that the stages worked for grief, but it wasn't something that she wanted to tackle until the end of her life." When I asked Kessler whether Kübler-Ross had done any additional research on grief, he replied, "She didn't make a distinction between one's own dying and grieving the loss of someone else, because dying is grieving itself. It's grieving the life you're never going to have. She saw them as fluid."

I had called Kessler to get his reaction to the news that a group of researchers at Yale University had decided to test whether the stages do, in fact, reflect the experience of grief. In the Kübler-Ross model, acceptance, which she defined as recognizing that your loved one is permanently gone, is the last and final stage. But the resulting study, published in the *Journal of the American Medical Association,* found that most respondents accepted the death of a loved one from the very beginning. The researchers interviewed 233 people between one to twenty-four months after the death of their spouses by natural causes to assess their "grief indicators," and across all points on the timeline acceptance was the indicator most frequently checked off. "Most bereaved individuals are capable of accepting the reality of the loss even initially," says Holly Prigerson, co-author of the study and now the director of Psycho-Oncology Research, Psychosocial Oncology and Palliative Care at the Dana-Farber Cancer Institute. On top of that, participants reported feeling more yearning for their loved ones—a condition researchers called pining—than either anger or depression, perhaps the two cornerstone stages in the Kübler-Ross model. "What might explain the sustained, widespread and uncritical endorsement of the stage theory of grief? From a human interest perspective, it may reflect a desire to make sense of how the mind comes to accept events and circumstances that it finds wholly unacceptable," Prigerson wrote in a subsequent editorial in the *British Journal of Psychiatry* in 2008. "Results from our study, together with enduring popular and scientific interest in the topic, suggest that it may be time to reevaluate stage theories of grief and consider their potential clinical utility." (Kessler told me he had not heard of Prigerson's study.)

Skepticism of the stages has been building steadily since the early 1970s, when Richard Schulz, then a twenty-four-year-old grad student in social psychology at Duke University, and his adviser, David Aderman, looked into the existing research to see if there was any support for the stages, which there wasn't. "As fairly hard-nosed scientists, we wanted to set the record straight by looking closely at popular ideas on death and dying," recalled Schulz, who is now a professor of psychiatry and director of the Center for Social and Urban Research at the University of Pittsburgh. Thirty years later, however, the stages still hold sway with professionals and lay people—a 2008 survey of fifty hospices in Canada found that Kübler-Ross's work was the literature most frequently consulted and distributed to families of dying patients, used by 75 percent of all respondents. When I asked Schulz why the stages seemed so resistant to debunking, he replied, "Because they have great intuitive appeal, and it's easy to come up with examples that fit the theory."

Kübler-Ross defenders say that she never intended her stages to be taken quite so literally ("it's just a theory") and that she herself warned that they don't always happen in sequence. But their inculcation shows just how powerful theories can be, and Kübler-Ross herself frequently referred to them as if they were established fact, and not untested hypothesis. It's not all Kübler-Ross's fault. We are to blame too for embracing a doctrine that, as I will examine in Chapter 2, has actually lengthened the expected duration of grief and made us more judgmental of those who stray from the designated path. We have been misled by the concept that grief is a series of steps that ultimately deposit us at a psychological finish line, even while social science increasingly indicates that it's more a grab bag of symptoms that come and go and, eventually, simply lift. "Stage theories of grief have become popular and embedded in curricula, textbooks, popular entertainment, and media because they offer predictability and a sense of manageability of the powerful emotions associated with bereavement and loss," says Janice Genevro, a psychologist who was commissioned by the Center for the Advancement of Health to do a report on the quality of grief services, and concluded that practitioners' techniques were misaligned with the latest research.

And so, when someone we love dies, we continue to grapple with a model for grief that's not only inaccurate but, at times, even punishing. Valerie Frankel, a novelist, recalls that in 2000, when she lost her husband, Glenn Rosenberg, to lung cancer, she found the stages (which she had "known about for forever, it's just standard knowledge") to be of no relevance at all.

"I simply felt depression," she says. "I wasn't angry at God about Glenn's death, although I did fly into a rage about something stupid that the doctors said. I don't think I did any of that bargaining stuff. And as for acceptance, well, you don't really have a choice." But it wasn't just stage theory that Frankel found misleading. Her father, a doctor, told her that it takes at least six months to two years for a person to recover from such a tragedy, and one of the books she read advised against starting any new romances for at least a year because her emotions were too unstable and might lead her into inappropriate or unhealthy relationships. After seven months, however, she joined an online matchmaking site and met a man named Stephen Quint. "We started having a lot of fun together and it was really life-affirming," she says. Six months after their first date, they got engaged, although Valerie says that despite the rapidity of their courtship, "It's not like [her husband Glenn's death] magically disappeared. Steve was really fantastic about understanding the whole transition." She is now happily remarried and says that, contrary to everything she'd heard and read about widowhood, beginning a relationship with Stephen so soon after the death of her husband was a stroke of good timing.

Compared to the way widowhood is typically portrayed, Frankel probably sounds like an unusual case. A six-month recovery window was thought to be unrealistic, until recent research conducted by George Bonanno, a psychology professor at Columbia University Teachers College, showed that it was more the norm than the exception. Bonanno has laid bare many assumptions about bereavement by following groups over long periods of time and using standardized questionnaires to measure their reactions (as opposed to Kübler-Ross, who spoke to her subjects once and asked open-ended questions). Bonanno and his colleagues tracked elderly people whose spouses died of natural causes, and the single largest group—about 45 percent—showed no signs of shock, despair, anxiety, or intrusive thoughts six months after their loss. Subjects were also screened for classic symptoms of depression, such as lethargy, sleeplessness, anhedonia, and problems in appetite, and came up clean on those as well. That didn't mean that they didn't still miss or think about their spouses, but by about half a year after their husbands and wives had died, they had returned to normal functioning, contradicting the often repeated saying about widowhood that "the second year is harder than the first." A much smaller group—only about 15 percent—were still having problems at eighteen months. An even smaller group, about 10 percent, exhibited a "recovery" pattern with grief symptoms moderately high about six months after the loss but almost completely gone by eighteen months. In addition, some respondents fell into two additional groups—people who were depressed before and after their loss whose troubles seemed to be a pre-existing condition, and people whose depression improved following the loss, suggesting that the death of their spouse actually relieved stress instead of causing it.

Many Americans who lose a loved one are more resilient than we give them credit for. The dominant grief culture in America today asserts that it's perfectly normal to get mired in a long and protracted reaction, when in fact this happens to only a small minority whose debilitating symptoms last considerably longer than six months and who might be suffering from a syndrome clinicians are now starting to call Prolonged Grief Disorder. (As I discuss later in the book, this subset is the only group that seems to be helped by grief counseling—the rest of the population does just as well on its own without it.) Our grief culture also defines grief as a project that must be actively tackled by identifying and vocalizing one's darkest feelings. The opposite may actually be true—one of George Bonanno's studies found that recently bereaved individuals who did *not* express their negative emotions had fewer health problems and complaints than those who did, suggesting that damping them down might actually have a protective function. Our grief culture maintains that "everyone's grief is unique," and then offers a uniform set of instructions. In fact, while researchers haven't come up with a universal description for grief (and in all likelihood, they never will), they have identified specific patterns to its intensity and duration. And while there are many factors that may make bereavement harder on some than others (such as the suddenness or cause of death, or the age of and relationship to the person who died), probably the most accurate predictors of how someone will grieve are their personality and temperament before the loss. Back in 1961, Edgar N. Jackson, a Methodist minister and popular author, suggested as much when he wrote the following in a little guide called *You and Your Grief:* "If one has always met life's problems with strength and assurance, it is reasonable to assume that he will meet this experience the same way. One who has been easily distressed by circumstances may be so disturbed by the encounter with death that he will need guidance and special help." Today, that kind of relativism is anathema. Instead, grief is portrayed as an abstract state that uniformly descends upon us.

Although I have lost people dear to me in my own life, this book did not grow out of personal experience but rather a journalistic desire to understand how we arrived at certain norms that don't seem to be serving us particularly well. In contemporary America, mourning conventions such as wearing black armbands or using black-bordered stationery have mostly disappeared, but they have been replaced by conventions for grief, which are more restrictive in that they dictate not what a person wears or does in public but his or her inner emotional state. Since these rules use a psychological model, they have an empirical gloss, when in fact they are largely myths, or, to borrow a term from two pioneers in debunking those myths, "clinical lore" that misinforms practitioners and the general public. The bigger question, one that I will try to answer in this book, is why we continue to look at grief through such a distorted lens.

My intention is not to diminish grief, which is painful and must be respected as such, but to reframe it in a way that may ultimately be liberating, both for those who have yet to face it and those who are currently in its throes.

1

1

The American Way of Grief

"Mourning never really ends, only as time goes on, as we do our work, it may erupt less frequently," said the man in an olive green suit, addressing a group of social workers, nurses, and hospice employees gathered in the public library in Cherry Hill, New Jersey. The man's name was Alan Wolfelt, described in his brochures as "an internationally noted author, teacher and grief counselor," and he gave the distinct impression that he had said those lines before. In fact, probably at least eighty times a year—that's how frequently he lectures all over the country. Every fall, Wolfelt embarks on his yearly speaking tour, flying to various cities from his base of operations in Colorado with an itinerary so hopscotch that if it were diagrammed it would resemble one of those flight route maps from the back of an airline magazine.

When Wolfelt's not on a plane or lecturing to caregivers, he can be found at the hexagonal-shaped Center for Loss and Life Transition in Fort Collins, which he designed, built, and opened in 1983. "I found my calling at age sixteen, when I wrote my mission statement that I wanted to start a center," he said, explaining that the death of a friend from childhood leukemia was the triggering event. Had Wolfelt come of age in the 1950s, that early experience might have moved him to become a doctor, or perhaps a clergyman. But Wolfelt happened to reach adulthood in the 1970s in the midst of a burgeoning movement devoted to the study of death and its impact on those left behind. The fulcrum of that movement, Elisabeth Kübler-Ross, was lecturing across the country, classes on death and dying were sprouting up on college campuses, and the term "thanatologist" for those who specialized in such studies was coming into widespread use. Wolfelt caught the wave and has ridden it ever since.

As a college student at Ball State University, Wolfelt spent three summers participating in Elisabeth Kübler-Ross's retreats, where, among other things, the bereaved were encouraged to scream and pound on telephone books and otherwise externalize their emotions. He wrote his master's thesis on children and death, and went on to earn a Ph.D. in counseling psychology, although he distances himself from the medical model in that he advocates "companioning" people who are grieving rather than treating them. (For $775, grief counselors can attend a three-day seminar to learn more about the difference between treating and companioning.) Wolfelt has written dozens of books—the exact number is hard to ascertain as at least three new volumes seem to appear every year. He also produces pamphlets and packets on grief for hospices and funeral homes, all through his own publishing company, whose sixteen-page catalogue is titled "The Writings of Dr. Alan Wolfelt." He has even written a "Mourner's Bill of Rights," a ten-point declaration printed on small cards ($15 for a packet of fifty) to keep in your wallet or hand out to other people, reminding them and yourself of the following:

1. You have the right to experience your own unique grief. No one else will grieve in exactly the same way you do. So, when you turn to others for help, don't allow them to tell you what you should or should not be feeling.

2. You have the right to talk about your grief. Talking about your grief will help you heal. Seek out others who will allow you to talk as much as you want, as often as you want about your grief.

And on it goes, up until the last:

10. You have the right to move toward your grief and heal. Reconciling your grief will not happen quickly. Remember, grief is a process, not an event. Be patient and tolerant with yourself and avoid people who are impatient and intolerant with you. Neither you nor those around you must forget that the death of someone loved changes your life forever.

In any other country, Alan Wolfelt's cottage industry wouldn't exist. There are grief counselors in other parts of the world, particularly in England and Australia, two countries that share many ethnic, religious, and cultural similarities with the United States. But in the United Kingdom, almost all grief counseling is administered through Cruse Bereavement Care, a charity founded in 1959 by a former Citizen's Advice Bureau worker named Margaret Torrie, whose original intent was to help widows handle practical matters such as housing, insurance, pension, and taxes. Cruse Bereavement Care (the name is a biblical reference to a widow's cruse, or jar of oil, which never runs out) trains thousands of volunteers to give advice and support for free to anyone seeking it. (As of 2005, Cruse had 5,400 volunteers backed by a paid staff of about 120, and both a central office and 240 local service offices. That same year, Cruse responded to 177,452 inquiries, representing approximately one third of registered deaths.) Australia's grief counseling is far less centralized—there are several regional organizations, but they are also mostly nonprofit and partner with or receive funding from state governments.

Although Alan Wolfelt may be America's most industrious grief advisor, he's certainly not the only one. There's also Therese Rando, author of the popular *How to Go On Living When Someone You Love Dies,* or Brook Noel and Pamela D. Blair, authors of *I Wasn't Ready to Say Goodbye.* Popular self-help gurus such as Melody Beattie (of *Codependent No More* fame) and Jack Canfield (the *Chicken Soup* guy) have expanded their own franchises into grief. But no one seems to have blanketed the market quite like Alan Wolfelt, offering grief-related products at every price point and carving out thinner and thinner slices on the same theme. He has written on suicide grief, holiday grief, workplace grief, teen grief, even pet grief. When I attended one of his lectures in the fall of 2009, he was promoting a book on grief people may be experiencing without even knowing it that was mysteriously titled *Living in the Shadow of the Ghosts of Grief* (do ghosts have shadows?). In the book, Wolfelt argues that a whole host of problems (depression, anxiety, bad relationships, general malaise) might be due to what he calls "hidden grief" or "carried grief" from a past loss that was driven underground but remains as toxic as a chemical spill. "It's an epidemic in this country," he asserted in his lecture. "We are a mourning-avoidant culture."

Almost every grief specialist out there makes a similar claim: that our society rushes grief or ignores it altogether, although they rarely cite any supporting polls or surveys. Usually, the closest they come to offering evidence is to point out, as Wolfelt did, that people are given only three days off from work for bereavement leave, a grossly inadequate amount of time to "process the loss." I heard this example used many times and eventually came to realize that it was a straw man argument that tells us more about how our country values work over family and leisure time than it does about our lack of sensitivity to grief. American workers get a notoriously low number of paid vacation days every year compared to other industrialized countries—an average of thirteen days, compared to twenty-five in Japan or thirty-five in Germany—so three days off when someone dies is certainly on scale. Meanwhile, maternity leave in the United States is usually only about six weeks to three months, a mere sliver of the amount of time it takes to actually raise a child from infancy to school age.

Nonetheless, the idea that death and, by extension, grief are ignored runs consistently through modern grief literature, from Kübler-Ross ("We live in a very peculiar, death-denying society," she testified before the Senate's Special Committee on Aging in 1972) up to contemporary observers. In 2006, poet and literary critic Sandra Gilbert argued in *Death's Door,* the most recent survey of the grief landscape, that, "Just as we've relegated the dying to social margins (hospitals, nursing homes, hospices), so too we've sequestered death's twins—grief and mourning—because they all too often constitute unnerving, in some cases, indeed, embarrassing reminders of the death whose ugly materiality we not only want to hide but seek to flee."

This argument has become so popular that it continues to get perpetuated even by those who seem to illustrate the opposite. In 2010, Meghan O'Rourke wrote in *The New Yorker* that after her mother died, her friends seemed ill at ease with her grief. "Some sent flowers but did not call for weeks. Others sent well-meaning emails a week or so later, saying they hoped I was well, or asking me to let them know 'if there is anything I can do to help.' " O'Rourke was not comforted by these platitudes. "Without rituals to follow (or to invite my friends to follow), I felt abandoned, adrift," she continued, without mentioning whether or not her family had held a funeral or memorial service, one ritual very much alive where close and old friends can show support by attending. O'Rourke finally made her way to this pronouncement: "In the wake of the AIDS crisis and then 9/11, the conversation about death in the United States has grown more open. Yet we still think of mourning as something to be done privately." If that's the case, then O'Rourke's author's note was certainly perplexing, as it announced that she is writing a whole book about her grief experience, which will join numerous other memoirs of loss published in the last decade that, while heartfelt, belie the argument that grief is private. As British sociologist Tony Walter has pointed out, when something is repeatedly characterized as taboo, as grief has been for the last fifty-odd years, that's a good indication that it is actually anything but.

This is not the first time in American history that the expression of grief has become so visible. Grief entered the public realm around 1865, where it remained for the next several decades; it then receded from view from about 1915 up through the 1960s, before surging once more up until today where we seem to have hit a new peak.

Our current public grief culture has its parallel in what is usually characterized as the Victorian period of the mid- to late nineteenth century. The Civil War had recently brought mortality into sharp focus—with casualties surpassing 600,000, or 2 percent of the population, almost every household in the South lost a father, brother, or son and every household in the North knew of one that did. The war was then capped by the murder of Abraham Lincoln, the first assassination of a sitting president in our nation's history, and the subsequent trial and public hanging of those who conspired with John Wilkes Booth in his larger plan to take down the entire government. Lincoln was given elaborate funeral processions in Washington, D.C., and New York City before his body was carried by train to his home in Springfield, Illinois, where it met with great crowds at every stop. In a eulogy to Lincoln, the clergyman Henry Ward Beecher (brother to Harriet Beecher Stowe) said, "No monument will ever equal the universal, spontaneous and sublime sorrow that in a moment swept down lines and parties, and covered up animosities, and in an hour brought a divided unity of grief and indivisible fellowship of anguish." Lincoln's death continued to reverberate for many years. Walt Whitman wrote several famous poems about it, including "When Lilacs Last in the Dooryard Bloom'd" and "O Captain! My Captain!" and was asked to recite the latter so frequently at speaking engagements up until the 1880s that he almost wished that he'd never written it.

It was during this period that mourning customs became much more elaborate, in part due to the standard set by England's Queen Victoria, who famously dressed in black for the rest of her life following the death of her husband, Prince Albert, in 1861. Americans as well as the British followed her lead, lengthening the time span for mourning and introducing bereavement-specific attire and accessories, although these customs were practiced mainly by those who had the money and leisure time to support them. As the historian Thomas J. Schlereth notes in *Victorian America: Transformations in Everyday Life,* "By the 1880s, a rigorous and detailed system of rules governed proper mourning dress and behavior. Women in 'full' or 'deep' mourning wore dresses of black bombazine and mourning bonnets with long, thick, black crepe veils." There was even mourning jewelry made out of black jet beads, and bracelets or watch chains woven from strands of the deceased's hair. Meanwhile, the duration of mourning was explicitly delineated according to one's relationship to the deceased, and always required withdrawal from society. According to an etiquette book published in 1887, "For one year no formal visiting is undertaken, nor is there any gayety in the household. Black is often worn for a husband or wife two years, for parents one year, and for brothers and sisters one year; a heavy black is lightened after that period."

In the late nineteenth century, death was in the process of being secularized—and to a certain extent, commodified. Instead of being buried by family members in churchyards, the dead were taken to new large, parklike cemeteries built outside the centers of cities. The funeral home, a new enterprise, began providing all arrangements, from retrieving, transporting, and preparing the body for burial. Embalming, which had been developed by chemists in part to preserve the remains of those killed in combat in the Civil War until they could be delivered back to their families, became more common, enabling the body to be kept on view for several days and giving birth to the tradition of the wake. Photography, also a recent invention, was put to use to commemorate the dead. Relatives often gathered around the deceased lying in his or her coffin or bed for one last family photo, and such postmortem images were often displayed on mantelpieces or pianos alongside other family portraits. The infant mortality rate was still high, and mothers were often photographed holding their dead child. As Schlereth points out, "The Victorian way of death fostered both sentimentality and science, emotion and expertise, as it transformed a traditionally brief, personal, private drama of the everyday life cycle into a prolonged, ceremonial, public ritual."

These changes coincided with a romantic movement in

the arts that encouraged the expression of love and loss. Grief as an emotion began to be depicted in plays and in paintings, where deathbed scenes became popular. There was also a new phenomenon of "consolation literature," manuals and hymns specifically for the bereaved, or anthologies of elegies for infants collected under titles such as *Tears for Little Ones.* One of the most popular novels of the day, *The Gates Ajar,* by Elizabeth Stuart Phelps, was the fictionalized first-person diary of a young woman whose brother died in the Civil War. "One week; only one week today, this twenty-first of February," the novel opens. "I have been sitting here in the dark and thinking about it, till it seems so horribly long and so horribly short; it has been such a week to live through, and it is such a small part of the weeks that must be lived through." Published in 1868, *The Gates Ajar* plumbs its protagonist's interior life as she chronicles her despair as well as annoying visits from neighbors and the town deacon. (*The Gates Ajar* was rereleased by the University of Michigan in 2005.)

Since women were the keepers of the hearth and home, they became "the chief agents for emotions at death, with men's responses more variable, but whole families were frequently engulfed in sorrow," wrote Peter N. Stearns in *Revolutions in Sorrow: The American Experience of Death in Global Perspective.* "Images of mourning abounded in popular art. The theme of grief, like death itself, seemed increasingly ubiquitous. The emotion was often hauled out for discussion and praise in nineteenth century family manuals. Grief was held to be the counterpoint of love, which in turn was now urged as the foundation for proper family life." This new romanticism meant that grief also became a barometer of the strength of someone's affections and loyalty. As Mrs. John Sherwood wrote in *Manners and Social Usages* in 1884, "If one did not mourn well, that is if one did not display grief in every acceptable mode, this demonstrated a lack of respect for the deceased . . . to go to the opera, or a dinner or a party, before six months have elapsed, is considered heartless and disrespectful." The injunction to grieve was in stark contrast to the Puritan ethos of pre-Victorian American life, when too much grief was seen as challenging the will of God, who alone knew when it was someone's time to leave the earth.

All of these forces transformed grief from a commonplace eventuality into an experience to be feared and to a certain extent sentimentalized. The newly intense emotion even came under examination by Sigmund Freud, who wrote his famous paper "Mourning and Melancholia" in 1915. This short paper was really more about defining clinical depression, but it contained the seeds of our current grief culture in this one sentence: "The inhibition and loss of interest are fully accounted for by the work of mourning in which the ego is absorbed." That "work," Freud conjectured, was for the ego to detach itself from the deceased (or in his sexual worldview, for the libido to withdraw from the love object) so that it could reattach itself to someone else. Later on, in the late 1960s, Freud's definition of grief as work became the guiding metaphor for modern grief theory, even though, as it turned out, the assumption that people need to tackle their grief like so much paperwork in an inbox, or the "grief work hypothesis" as it is now called by a growing number of skeptics, was an unsubstantiated one.

The Victorian period of grief came to a close when World War I forced Americans to accept death proudly and unemotionally so as not to undermine the war effort. But even before then, the emotional displays of the era had already started to come under attack. In an article published in 1911 in *Harper's Bazaar,* "Facing Death," author Louise B. Willcox argued that the customs of her day had become indulgent. "Grief is self-pity," she wrote. "Perhaps if we were less centered upon our own happiness, grief over the loss of our beloved ones would not be the terrible thing that it is."

In England, extended mourning became impractical due to the sheer number of casualties from World War I. Restraining rather than expressing one's despair became the new protocol, and those who bore their losses quietly were held up for praise. An article published in 1916 in the American magazine *The Literary Digest* commended the new mien of grief of the Europeans, "those millions bereaved by the present war [who] have need of words of comfort, especially since theirs is the duty not only to endure but to efface as far as possible the signs of woe." The article went on to compliment the "quiet heroism and endurance" of those who have "set up a new and noble precedent in the matter of courage and self-control.

Out from their secret chambers they come, with washen face and brave lips to do their duty and refrain themselves. How beautiful it is! What a fine thing to see!" The United States didn't enter the war until 1917, and therefore suffered fewer casualties than Great Britain and France, but family members of American soldiers who were killed were similarly encouraged to feel pride in their heroic sacrifices.

Visual representation of death faded and postmortem photography fell out of favor. Grief manuals became few and far between, and those that did appear seemed to espouse an unsympathetic, move-on-with-your-life pragmatism. In 1941, just as America was entering another world war, a widow named Toni Torrey came out with an advice book in which she suggested that women who lose husbands should not expect to be plunged into prolonged grief, nor be alarmed if they're not: "Exclusive and important as you may feel in your absorbing grief, your experience is not unique. A great many widows have traversed the road before you—a great many more will follow," she chided her readers. "You don't have to be reminded, of course, that you are living in a cellophane world. People can look through and see the real you. A parade of grief is seen through, too. It is more than ever unnecessary."

By the 1950s, intellectuals began remarking that death in contemporary society had become all but invisible. In 1955, the British historian Geoffrey Gorer wrote a famous article, "The Pornography of Death," in which he argued that death was now more unmentionable than sex. "The natural processes of corruption and decay have become disgusting, as disgusting as the natural processes of birth and copulation were a century ago," he noted. Gorer attributed this shift to several factors: the first was that public health measures and advances in medicine had made natural death, especially among the younger segment of the population, relatively uncommon in home life. The second was a decline in the number of people who believed in the afterlife, even among religious observers. "Without some such belief, natural death and physical decomposition have become too horrible to contemplate or to discuss," Gorer suggested. Prominent French philosopher Philippe Ariès suggested a third reason why Americans seemed particularly averse to death: it interfered with one of our inalienable rights, our pursuit of happiness. "One must avoid—no longer for the sake of the dying person, but for society's sake, for the sake of those close to the dying person—the disturbance and the overly strong and unbearable emotion caused by the ugliness of dying and by the very presence of death in the midst of a happy life, for it is henceforth given that life is always happy or should always seem to be so."

Humanist psychologists such as Erich Fromm and Rollo May put our national personality on the couch and found it cold and profoundly alienated. In his 1955 book *The Sane Society,* Fromm wrote, "Instead of allowing the awareness of death and suffering to become one of the strongest incentives for life . . . the individual is forced to repress it. But, as is always the case with repression, by being removed from sight the repressed elements do not cease to exist. Thus the fear of death lives an illegitimate existence among us. It is one source of the flatness of other experiences, of the restlessness pervading life, and it explains, I would venture to say, the exorbitant amount of money this nation pays for its funerals." Jessica Mitford's critique of the funeral industry in *The American Way of Death* in 1963 was a wake-up call that there was something both superficial and delusional about how we honored our loved ones.

Passionate grief, although without the elaborate clothes and rules for mourning, was poised to surge again: this time, it was the assassinations of JFK, Martin Luther King Jr., and Bobby Kennedy that brought mortality into sharp focus, although Jackie Kennedy's lack of tears at her husband's funeral in 1963—among the earliest to gain a widespread audience through television—was an example of the era's restraint. That same year, the first regular course on death was introduced at an American university, and interest in the topic gradually grew. But it was the publication of Kübler-Ross's book in 1969 that opened grief's floodgates once again.

Demographically, the 1970s were the perfect time for a return to emotionality. Baby boomers were rejecting the detached ways of their elders and learning about their psychic life. Even those who were not in therapy were turning inward thanks to Gail Sheehy's *Passages,* a bestseller based on a UCLA psychiatrist's concept of "adult life stages," which included a growing awareness of mortality between the ages of thirty-four and forty-five. Self-awareness, and its articulation, became the new national prerogative. People were looking for answers, and one of the questions they had was, How is death going to affect me?

In 1971, Jane Brody of *The New York Times* wrote of a "growing league of professionals that is trying to throw new light on a long-taboo subject that sooner or later touches every human being." Stand-alone centers for grief began popping up in cities across the country. In 1974, the memoir *Widow* by Lynn Caine became a bestseller and was later made into a TV movie. The review in *The New York Times* praised it as a "common-sense guide for mourning" in which Caine told of "blundering blindly through the stages of grief which psychiatric studies have shown to be inevitable and predictable." (In fact, there were no such studies.) The same year, *Publishers Weekly* announced, "Death's now selling books." In 1975, the country was polarized by the case of Karen Ann Quinlan, the Terri Schiavo of her day, whose parents fought a court battle to take their comatose daughter off a ventilator. In 1977, a play about three terminally ill patients called *The Shadow Box* won a Tony Award and the Pulitzer Prize for Drama, and was quickly adapted into a TV movie directed by Paul Newman.

Death and grief flourished as areas of academic study as

well. In 1976, Robert Fulton, who had founded the Center for Death Education at the University of Minnesota, estimated that more material on death and dying had appeared in scholarly journals between 1970 and 1975 than had appeared in the previous one hundred years. The movement had a few detractors: in an editorial published in *JAMA* in 1974, cardiologist Samuel Vaisrub deplored the trend, saying, "It is much too easy to write about death. There is no need for statistical evidence and for lengthy references." In 1976, scholars and counselors founded the Association for Death Education and Counseling, or ADEC, and began publishing a journal, *Death Studies,* which exists to this day. (In the debut issue, published 1977, almost every article was either about counseling the dying or educating others about their needs; today, the majority of a typical issue is about those who are struggling with someone else's death.)

This shift in focus from the deceased to the griever coincided with a radical change in the way society itself publicly mourned its biggest losses from symbolizing the achievements of those who had been sacrificed to re-creating the sacrifice itself. During the 1980s, two large groups of Americans were recognized in this new way: those who died in the unpopular Vietnam War, and those who died of the frightening new disease of AIDS. Unlike earlier memorials where one symbol, usually a statue of a human figure, is used to represent the whole group, these memorials honored each dead person individually and by name. In doing so, they evoked not so much abstract principles such as bravery or sacrifice but the cumulative toll of the losses themselves.

As anyone who has visited the site knows, walking down into the Vietnam Veterans Memorial in Washington, D.C., is an unsettling experience even for people who don't know anyone who died in that war. Maya Lin's design of two long black triangular walls descending into the earth to meet perpendicularly in a tall corner that obstructs your view of anything else was intended to evoke a giant crypt containing every one of more than 58,000 names etched into the granite. "We, the living, are brought to a concrete realization of these deaths," wrote Lin, then a young architecture student taking a funerary class, in her original, one-page submission to the design competition. "Brought to a sharp awareness of such a loss, it is up to the individual to resolve or come to terms with this loss . . . the area contained within this memorial is a quiet place meant for personal reflection and private reckoning." Since its unveiling in 1982, the memorial has in fact become something of a shrine, with people leaving photos and flowers and all kinds of offerings (which are collected and stored by the National Park Service), and taking away rubbings made from the names on the wall.

In 1987, gay rights activist Cleve Jones created the first panel of what would become the AIDS Memorial Quilt and formally organized the NAMES Project Foundation to fund it. By the time the Quilt made its debut on the National Mall in Washington later that year, it was already larger than a football field and was visited by half a million people, an overwhelming response considering the stigma the disease carried at the time. Both the Wall (as the Vietnam Veterans Memorial is now called) and the Quilt also foreshadowed how those who died in 9/11 would become known individually, first through the thousands of missing persons flyers posted around New York City, then in the media, especially the "Profiles of Grief" series in *The New York Times,* and finally with a reading of the names on a nationally broadcast ceremony every anniversary.

When the wars in Iraq and Afghanistan were under way, the Bush administration tried to manage this new individuation of loss by upholding a ban on photographs of soldiers' flag-draped caskets as they returned back to the United States. President George H. W. Bush had initiated the ban during the first Gulf War, when the practice of returning dead soldiers to U.S. soil was still a recent development: Until the Korean War, bodies of U.S. servicemen were often buried overseas in one of twenty-five military cemeteries in France, England, and Mexico that serve as the final resting place for almost 125,000 soldiers. Shortly after Barack Obama took office, Defense Secretary Robert M. Gates reversed the ban, announcing that the media would be allowed to take pictures of the coffins as long as family members agreed.

Bereaved families of military personnel became increasingly vocal about their losses, and their grief fueled critiques of our military engagement. "While during the Vietnam War grieving families were silenced by the anti-war protest movement, in this war the faces of grieving mothers—from Michael Moore's portrayal of Lila Lipscombe grieving for her son in *Fahrenheit 9/11* to Rose Gentle and Cindy Sheehan making public their private grief over the deaths of their sons in Iraq as a form of protest—have become central to the discourses surrounding the war and its conduct," noted Professor Carol Acton of the University of Waterloo in Canada.

Instead of hiding behind closed doors, war widows invited observers into their homes. Jim Sheeler, a reporter at *The Rocky Mountain News,* even spent a year following a "casualty assistance calls officer" whose duty it was to notify the next of kin when a Marine was killed, and his subsequent series won a Pulitzer Prize and was published in book form with photographs. In 2007, after her husband was killed in Iraq by a roadside bomb, Taryn Davis made a documentary about six other military widows, then started a Web site and national support organization called the American Widow Project.

This is not to say that attention shouldn't be drawn to the lives lost in Afghanistan and Iraq, which as of this writing numbered more than 5,600 Americans (and untold thousands more Iraqis). On the contrary, all those fatalities deserve to be recognized and accounted for.

But as grief has become increasingly visible, the standards

and codes of our emotional behavior have become more circumscribed. What began as a legitimate reaction against the impersonality of modern death has hardened into a doctrine that dictates not just our reactions but how we define the experience—to ourselves and to the rest of the world. We see it in medicine, in support groups sponsored by hospitals, and the form letters sent out at regular intervals to families from bereavement coordinators. ("Many people find this period of time, about six months after a significant loss, more stressful than they had expected," reads one letter from Massachusetts General Hospital Palliative Care Service, which recommends reading *Tuesdays with Morrie* and watching the *Chicken Soup for the Soul Live!* DVD.) We see it at funeral homes, which offer "aftercare services" to bereaved customers along with "pre-planning" your own funeral because, as material from the National Funeral Directors Association asserts, one benefit to doing so is "alleviating the responsibility from your family for making these kinds of decisions during an already difficult time of grief." We see it in the new ways that we publicly memorialize our dead, with wall murals and R.I.P. decals on the back windows of cars, or "ghost bikes" erected at the site of vehicular homicides. (Art historian Erika Lee Doss calls the steep rise in the number of both official and unofficial monuments "memorial mania.") We see it in new specialized grief organizations, such as Wings of Light (for people who lose loved ones in airplane accidents) and Sibling Survivors of Suicide. We see it online at Legacy.com, "Where

Life Stories Live On" through $49 Web pages for the dead, or its competitor, Tributes.com, created by the founder of Monster.com with funding from the Dow Jones Company.

Even if the movement has enriched a few individuals, it is driven more by ideology than money. Grief counselors are, by and large, not a sinister bunch out to make a buck off of other people's misery, but they do, in the interest of self-preservation, have a stake in convincing us that grief is long, hard, and requires their help. One unfortunate result of this bit of mythmaking is that our expectations of others have become more restrictive—which is paradoxical considering that awareness and tolerance were among the grief movement's primary goals. In a survey conducted in 1970, when asked whether and when a widow or widower should get re-married after the death of a spouse, 24 percent said that it was unimportant to wait, and 31.5 percent said it was okay within one year's time. Contrast that with data collected in 2000 from a replication of that same survey and you see that public opinion has shifted dramatically: a mere 6.7 percent said that it was unimportant to wait, and only 8.8 percent said it was okay to get married after one year. While we have become more aware of the diversity of romantic relationships—and relaxed our attitudes about interracial dating, cohabitation, divorce, homosexuality, and gay marriage—we have actually grown *more* conservative about remarriage after spousal loss. When asked why he thought his results showed an increase in the amount of time considered appropriate before getting married, Bert Hayslip, a professor of psychology at the University of North Texas who conducted the replication survey, said that it could "reflect a greater trend toward being judgmental, perhaps reflecting their own greater anxiety about losing someone."

Which leads us to a central question: Why would we be getting *more* anxious about losing someone, even while modern medicine continues to prolong people's lives? Our modern grief culture has created that anxiety by promoting two intertwining beliefs: (1) that grief is necessarily lengthy and debilitating; and (2) the only way out is to work through it—in a series of stages, steps, tasks, phases, passages, or needs. These two tenets have both been challenged by recent research, yet they are still unavoidable to anyone looking for guidance or information about grief.

2

2

Is Widowhood Forever?

It was supposed to be a good beach day—the forecast called for 74 degrees and sunny—and in Fort Lauderdale in January the weather is at the top of everyone's mind. At 9:30 in the morning, an unusual sight greeted tourists who were already staking out the best deck chairs on the waterfront patio of the Ocean Sky Hotel. On the beach below, about two dozen long-stemmed red roses had been arranged in the sand to form the shape of a heart. A group of people—some young, some middle-aged, mostly women—stood quietly in a circle around it.

A barrel-chested man in shorts and T-shirt motioned to the others to move in and hold hands. This was Ray Schmuelling, a former Air Force officer from Kentucky whose wife had died of ovarian cancer four and a half years earlier, at the age of forty-three. The night before, after a big group dinner at Nick's Italian Restaurant, Ray and the others had emptied a cooler of beer and wine, but the morning's event, a beach memorial service, had the group regarding him more somberly now.

"I've never done this before, so I don't really know what to say," he began. "But I was thinking that memorials aren't just about the person who died, but are also for the people they left behind, and the part of your life that died when they did. And I was also thinking that grief is like a journey through the waves, so it's fitting that we're doing this on the beach. Some of those waves are small, and some of them are so big that they knock you down, but they have to just keep coming before the water can get calm again." Ray picked up one of the long-stemmed roses and brought it to the water. The others followed suit, fanning out along the beach and facing the horizon line. Slowly, they began to release their roses into the ocean, some with an underhand toss, some scattering the petals as if they were ashes. Some of them hugged one another afterward in solidarity, while some stood by themselves.

Most of the people were meeting for the very first time, and many didn't even know one another's real names, only the tags they used online—HisBabe, 4boysmom, Country_ Gal. Theirs was a kinship forged out of tragedy, or as one attendee put it, "We all belong to a club that no one wants to join." They had flown here from points north and west for a gathering known as a Widowbago, organized by and for users of YoungWidow.org and held every week or so at locations nationwide, although most take place over lunch or dinner with an occasional long weekend such as the annual retreat in Fort Lauderdale. Ever since the Web site was launched in 2001 (by sheer coincidence, it went live after several months of planning on September 11), it has registered more than fourteen thousand members. Like many online communities, it has also created its own lexicon: widders or widdas (both gender-neutral terms referring to men or women); widow brain (spacey and forgetful); skin hunger (craving physical contact), and the dreaded DGIs (those unwidowed people who just Don't Get It). There are the "newbies" who post in the sections titled "Newly Widowed (1 Day to 6 Months)" and "Shock Wears Off, Reality Sets In (6 to 12 months)." Then there are veterans like Ray, who has been on the bulletin board for more than four years and attended his first Widowbago in 2005.

"I can't even go into the active [online] grieving rooms anymore because the emotions there are so raw," Ray said at dinner on the first night of the Florida gathering, in 2009. In addition to planning and attending the Fort Lauderdale weekend for the last four years, Ray also organizes a camping Widowbago on a lake in Tennessee every summer. "I still need the board because I need to figure out how to put my life back together, and I need it for social reasons," he said. His long-term involvement with the group raises the question of whether its function should be to see people through their intense grief by getting them to the point where they no longer participate, or whether, like alcoholism or surviving cancer, widowhood is forever.

Unlike with offline support groups, which usually last between eight and twelve weeks, no one gets automatically cycled out of YoungWidow.org. Members, some of whom stay active regardless of positive new turns in their romantic lives, say what keeps them coming back is that it's the only environment in which they feel completely understood and accepted, where they can crack jokes about organ donation one minute and cry on each other's shoulder the next (even if those shoulders are virtual). That feeling of solidarity traces back to the very first Widowbago, in January 2003, when a small group of the site's users decided to get together at the Crowne Plaza Hotel in New York City. One attendee joked that they should rent a Winnebago to roam the country and meet everyone from the Web site, and another said, "Then it would be a Widowbago!" and the term stuck.

Two widows from that original meeting were in attendance in Fort Lauderdale, though both were by now remarried, six years on. One of them, Patty, had met her new husband at a Widowbago. "They're sort of like hope for us," another woman, Sue, said of the remarried widowers. Sue joined the board in 2006 and now, a few years later, seemed torn between not wanting to be forever defined by her husband's death while at the same time wanting it never to be forgotten. She said she gets irritated when she fills out official forms at, say, a doctor's office, and is forced to check the box that says "single." Widows aren't really single, she said, "at least, not until we want to be."

If there is a third rail, a subject that is not supposed to be discussed, it is the possibility that grief may be finite. *There is no timeline for grief,* is how the advice books and Web sites put it. Even the concept of recovery itself is seen as a misleading, elusive goal. Though Kübler-Ross identified acceptance as her final stage, implying some kind of end point, she also said that you could never fully close the chapter on grief. "The reality is that you will grieve forever," she concluded in *On Grief and Grieving.* "You will not 'get over' the loss of a loved one; you will learn to live with it." This undoubtedly may be true for many, but the grief movement has taken that statement to mean that no one *should* ever get over such a loss, although that rule seems to get more strictly applied to women than men. (If a widow shows interest in dating or remarrying within the first year or two, it's often seen as a betrayal of her dead husband, while a widower is usually forgiven for such transgressions, perhaps because of the popular stereotype that a man can't possibly manage a household or look after children without a woman's touch; think of all of the TV sitcoms—from *The Andy Griffith Show* and *My Three Sons* to *The Courtship of Eddie's Father* and *Full House* and *Everwood*—that have been based on this premise.)

The last decade has been particularly demanding of widows. It began with 9/11, when Lisa Beamer, Lyz Glick, Kristin Breitwieser, and other women whose husbands had died in the terrorist attacks became public figures. As Susan Faludi pointed out in her book *The Terror Dream,* the 9/11 widows were sanctified in the press but also held to a strict code of behavior. As long as they adhered to the script of a devout wife and devastated survivor, they remained in good standing. When they deviated from it by demanding money or accountability from the federal government or by getting engaged or buying a new house or car, they were frequently demonized and recast as bad widows. Faludi quoted one firefighter's widow observing, "The public wants you to live up to what they made you. They don't really want you to move on."

In April 2002, Laura Mardovich made news as the first 9/11 widow to remarry. Her husband, Edward Mardovich, had been the president of a brokerage firm located in the South Tower of the World Trade Center. Despite the fact that her new fiancé was a fellow widower with children and a longtime family friend, the media portrayed her as dancing on her husband's grave. The *New York Post* reported ominously, "After Ed's death, Laura changed," and quoted an unnamed and unhappy relative: "I don't understand," she said. "I can see [Laura] wanting to go on with her life. I always wanted her to remarry, but I never thought she would do it this way." Comments from readers were more pointed. "If these two weren't having an affair before her husband's death, I'll eat my hat," read one. "This is akin to having a puppy run over and then just running down to the pet shop to get a new one," read another.

Even though the relative never explained what in particular she found most objectionable, it seems clear that Mardovich's crime was remarrying too soon. Thirty years ago, starting new romantic relationships within the first year after the death of a spouse was fairly commonplace. In 1977, the psychiatrist Stephen Shuchter founded the San Diego Widowhood Project and began tracking seventy people (twenty-one men and forty-nine women) from one month after their spouse's death until their fifth year of bereavement, interviewing them every three months for the first two years and then yearly. At ten months after the death, 67 percent said they were interested in dating and 33 percent had actually done so in the previous two weeks. What's more, 13 percent said that they were already living with someone of the opposite sex, and 6 percent had remarried. Today, good widows—one thinks of Mariane Pearl, whose reporter husband, Daniel, was held hostage and murdered, or Katie Couric, who started a colon cancer foundation in her husband's name—are fiercely protective of their husbands' legacies and tend not to get romantically involved with another man for a very long time.

Before Kübler-Ross's word became a cultural touchstone, the psychological distress of widowhood was not characterized as interminable. In 1964, Colin Murray Parkes, a psychiatrist at the Tavistock Institute of Human Relations in London, analyzed the medical records of forty-four widows and found a sharp rise in the number of psychiatric complaints in the first six months, followed by a return to a level similar to his control group of nonwidowed women, findings that he deemed were "consonant with the traditional picture of grief as a severe but self-limiting affective disorder."

George Bonanno's recent research confirms these early estimates of a limited time span for the worst of grief for the majority of people. Again, by his breakdown, the most common pattern showed no symptoms six months after the death, with smaller groups recovering by eighteen months or finally getting better between two and four years out, although never quite returning to normal. (Bonanno assessed his subjects for features of intense grief, such as depression, despair, anxiety, yearning, intrusive thoughts, and shock. According to other researchers, even while those core symptoms go away, people still continue to think about and miss their loved ones for decades. Loss is forever, but acute grief is not, a distinction that frequently gets blurred.) Bonanno's study followed mostly elderly couples, and one might expect that so-called natural deaths by illness or old age are easier to accept and recover from than more sudden or tragic deaths. But getting back on one's feet by about six months is certainly not the picture that you get from our most popular grief accounts.

Take, for example, Joan Didion's enormously successful

The Year of Magical Thinking, in which the writer describes being so shocked by the death of her husband, John Gregory Dunne (even though he was seventy-one and had a history of heart disease), that she continued to harbor hope that he was still alive for a full year. "Grief has no distance. Grief comes in waves, paroxysms, sudden apprehensions that weaken the knees and blind the eyes and obliterate the dailiness of life," Didion wrote. Compounding matters, Didion and Dunne's daughter, Quintana, lay ill in a coma at the time (and later died after *The Year of Magical Thinking* was published) so Didion had to postpone Dunne's funeral for several months. When I asked George Bonanno what he thought of *The Year of Magical Thinking,* he said, "The extreme reaction that Joan Didion describes is *not* the common reaction, but it's so unpleasant and horrible and poignant that it really captures our attention." (Other grief researchers ventured out on a diagnostic limb with Didion: "She describes elements that lead me to wonder if she is at risk for what we call Prolonged Grief Disorder, the biggest predictor of it being dependency on the deceased for a sense of role in life or identity," said researcher Holly Prigerson, who co-authored the study on Kübler-Ross's stages published in *JAMA* in 2007.) Another problem with first-person accounts such as Joan Didion's is that they often begin in medias res and don't usually offer a picture of the narrator's emotional baseline before his or her loved one died. Didion may well have suffered from depression long before her husband had his heart attack. Even if it had been treated and abated, any history of a mood disorder would make her more likely to experience a recurrence after a cascade of tragic events such as the death of her husband and fatal illness of her only child.

It's impossible to not be moved by Didion's account of this terrible chapter in her personal history. Indeed, *The Year of Magical Thinking* has become the standard to which all other grief memoirs will inevitably be held. What's troublesome about her approach, especially in the stage adaptation of the book performed by Vanessa Redgrave in both New York and London, is how Didion encourages her audience to see her extreme reaction as universal and archetypal. "This happened on December 30, 2003," the play begins. "That may seem like a while ago but it won't when it happens to you. And it will happen to you. The details will be different, but it will happen to you. That's what I'm here to tell you." As *The New Yorker* critic John Lahr, one of Didion's few detractors, pointed out in his review of the play, "In this misguided act of exhibitionism, [Didion] seems to imply that she's got a lock on loss. More important, by telling the viewers how to identify with her story, she robs them of her work of discovery. Mourning is about achieving separation . . . but the purpose of *The Year of Magical Thinking* is not to let go of Didion's lost loved ones . . . but to keep them alive. Far from signaling the end of her year of magical thinking, the play turns out to be an extension of it." The problem is not so much that Didion refuses to go gently into widowhood, but that she seems incapable of seeing a way out.

We don't usually come across books or plays or movies about women who have had the more typical reaction, women who begin to stabilize after about six months and start dating after a year because, to state the obvious, the more common trajectory is less interesting and less romantic. After the curtain fell at the performance of *The Year of Magical Thinking* that I attended in New York City, the ladies' room was packed with red-eyed women waiting in line to wash their faces and blow their noses. But having become accustomed, perhaps, to the more dramatic narrative, we have begun to expect all widows to adhere to it. And if they don't, we usually find a negative explanation for their recovery—that they didn't really love their husband, or were cold, unfeeling people, or were in denial and would eventually have a delayed reaction.

Knowing that these are common conclusions about widows who don't grieve for a long period of time, researcher George Bonanno asked his subjects about the quality of their marriages and found no significant differences between those who recovered quickly and those who took much longer. He found that by and large, the quick recoverers weren't blocking out thoughts from the past—in fact, they derived the most comfort from positive memories of their marriage. Nor were they found to be more cold, aloof, or distant when interacting with others. As to the possibility that they were repressing their grief, Bonanno followed the group for up to four years (some participants dropped out) to see if people who initially showed lower distress had delayed reactions, and found that none did. Across the board, symptoms seemed to decline over time, even for the people who took the longest to recover.

The danger here is that by normalizing an extreme reaction such as Didion's, we're inclined to pathologize a more normal reaction (and by normal I mean the statistical norm). "People think that widows are supposed to look a certain way—you're supposed to be in a dark corner in the fetal position, or abusing drugs or something," says Sarah White Bournakel, whose husband, Stefan Bournakel, a thirty-four-year-old optometrist, was killed on Maui in 2004 when an oncoming car swerved into his lane and forced him up an embankment. Sarah had met her husband in college and they had lived together for years before marrying in 2001. They had just moved to Hawaii, where Sarah gave birth to their first child, a boy named Nicos. Sarah remembers having to bring Nicos, then just one month old, into the morgue to identify her husband. "Stefan's hands were still in the driving position, his eyes were open and his body was warm and it was almost as if he had a smile on his face," she says. "He had only a little bit of blood trickling out of his ear but horrible internal injuries." Sarah put Nicos in his car seat at the base of Stefan's gurney so that she could hug her dead husband and cut off a lock of his hair. She alternated between nursing Nicos and tending to her husband's dead body for the next several hours, until a doctor told her there was a risk of hepatitis to the baby and she left.

In the months after Stefan's death, Sarah began to focus her energy on her own physical and mental health, but found that people tended to interpret it as disloyalty to Stefan. "I started working out with a trainer, I went to an acupuncturist and a nutritionist because I wanted to take care of myself, but the perception was 'She's on the party patrol.' I remember once I'd gotten my hair highlighted and someone said, 'Wow, you look great, you must not be that sad.' " As for her fellow widows, they saw her desire to move out of a place of deep grief as a sign of deep denial. Her sister, who had a master's degree in psychology, recommended that she find a widows support group, so she attended one at CorStone (formerly known as the Center for Attitudinal Healing) in Sausalito, California. "Their partners had passed away three, four, five years ago and Stefan had died only four months earlier," Sarah says. "An older woman said that she was just getting used to going to holiday events by herself. They came to me and said, *Sarah, do you want to share?* I said, 'I want to share the fact that I really don't want to be here in five years, life is too short, and so if you guys have any tips, I would like to hear them.' And they said, 'We understand, you're still in the shock phase.' "

Obviously, there is much variability, as not everyone grieves at the same pace. But research has shed new light on what make some people bounce back quickly and others prone to getting stuck. Resilient grievers appear better equipped to accept death as a fact of life and tend to have a more positive worldview. Chronic grievers seem less confident about their coping abilities and more dependent on the relationship to the deceased. (A lack of social support and financial resources also plays a role.) These differences become apparent within the first month and are good predictors of how someone will handle the loss over time, with early success seeming to set the course toward greater well-being, while early difficulties—intense negative emotions such as the desire to die or frequent crying—are associated with poor coping after two years. Researchers have also explored "attachment style"—how people react to being separated from a significant figure in their lives, thought to be determined by whether our parents were supportive and available, or narcissistic and aloof—and the bearing it has on grief. Canadian psychologist Tracey Waskowic surveyed seventy-seven widows and widowers ranging in age from forty-one to ninety-seven and found that securely attached people were less angry, less socially isolated, and less prone to guilt, despair, and rumination (which in itself is thought to perpetuate depressed moods) than insecurely attached people.

This is not to say that widowhood isn't painful, or that it doesn't require considerable adjustment. But it doesn't universally plunge people into a long-term state of devastation as it is so often portrayed. In 2003, using data from the Women's Health Initiative (a longitudinal national study with investigators at hospitals and universities across the country), researchers compared large groups of widowed women to married women and found that while recently widowed women reported more mental health problems than do married women, longer-term widows whose husbands had died over a year earlier actually reported *fewer* problems than did comparably aged married women, possibly reflecting the burden of having to care for sick husbands more than losing them. Nor, as it turns out, is widowhood an automatic recipe for loneliness, at least compared to other possible endings to a marriage. German psychologist Martin Pinquart surveyed over four thousand people aged fifty-three to seventy-nine and found that divorced men and women reported feeling *more* lonely than widows or widowers. (One reason may be that the divorced parents had less contact with their adult children, perhaps because divorce puts a strain on those relationships by forcing offspring to divide their time between their two parents.)

Then there is the question of what, exactly, creates the most problems for widows—is it the loss of a dearly loved companion or the stress, financial and otherwise, brought about by suddenly being single? Nowadays with pensions, life insurance, Social Security, and welfare, a woman who enjoyed financial security while married is rarely left destitute after her husband dies. The only current study I was able to find that addressed the financial changes of widowhood had somewhat inconclusive results. Investigators for the Institute for Research on Poverty looked at perceived financial happiness after widowhood, which, though not entirely reliable, is a better indicator than absolute financial status, since not all women might be distressed by a drop in income after widowhood as long as their basic needs were being met. Using data from 552 widows and widowers aged sixty-two to sixty-seven surveyed in 2004, the study found that being widowed had no effect on men's satisfaction regarding their financial status and only widows whose husbands had died between three to seven years earlier were less satisfied with their financial status. Women widowed less than three years or more than seven years did not report being less satisfied with their finances.

So where did our conception of widowhood as the worst thing that could happen to a woman come from? It seems to have been based on two surveys conducted back when women, especially older ones, *were* more dependent on their husbands, both for a sense of identity and financial security. Aside from now being very out of date considering the seismic changes that have taken place for women in the last four decades, the design of the studies themselves had major flaws.

The first was carried out in Boston from 1967 to 1973, after the director of the Laboratory of Community Psychiatry at Harvard Medical School recruited psychologist Phyllis Silverman to come up with an intervention program to prevent problems stemming from widowhood. Thinking that peer counseling would work best, Silverman recruited a dozen widow "aides" (women who had been widowed for a while), and over a period of two and a half years they contacted every new widow in one neighborhood in Boston, 430 women in total, and offered to talk. Even though Silverman eventually concluded that "very few widows ever seemed in danger of developing serious emotional illness," her subsequent report set a dire tone. "Since grief is unending, recovery is not possible," she wrote. The best a widow could hope for was to gradually adapt by moving through a series of stages. "At each stage, the individual has different needs and different tasks to accomplish," she wrote. "Moving from one stage to the next generally requires having completed some of the work of the preceding stage." Silverman proposed only three stages—impact, recoil, and accommodation—but cautioned that they occur slowly over time. As one widow's aide said, "It may take as long as three years to realize that you really are a widow."

In her analysis, Silverman argued that widowhood was so difficult because the death of a husband presented a massive identity crisis for women. "While men need others, their self-development focuses more on individuation and autonomy. A woman's identity is largely framed by relationships and is attached to the roles associated with these relationships. In losing an essential relationship, she loses an essential part of herself. Her grief is pervasive." Whether Silverman's conclusions were colored by gender stereotypes popular in psychology at the time, most of the women who responded to her outreach were, in fact, full-time homemakers when their husbands died. (Many of them did not even know how to drive.) It was the women who declined help from the aides, and therefore never became part of the study, who had worked outside the home before their husband's death and continued to work. Of them, Silverman noted, they were "correct in their appraisal" that they didn't need help when it was offered, although that fact did not change her final conclusion that widowhood universally presented major and long-lasting damage for women.

Despite the limitations of the sample, Silverman's Widow-to-Widow Project became the model for other bereavement support services. In 1974, the AARP launched the Widowed Persons Service, and the Community Mental Health Act of 1976 mandated that mental health centers receiving federal funds develop widow-to-widow programs as part of their education efforts. From 1981 to 1982, a sociologist named Julie Ann Wambach conducted a field study of three such peer support groups and noted in her write-up that "The grief process was accepted by the widows and professionals [her term for the widows who led the groups] as a fact that was not contestable. Most had never considered how the grief process had an origin, that it was a social creation to explain how survivors respond to the loss of a loved one. When pressed, the widows and professionals attributed the grief process to

Elisabeth Kübler-Ross. What they were referring to was not a grief process, but Kübler-Ross's stages of dying."

Wambach also noticed that there were certain unspoken rules about how one could talk about the death of one's husband. "I realized that if they said, 'I'm feeling okay,' it was seen as a betrayal," she recalls. "They had to find ways to not look callous." Wambach says that she was also amazed that every one of the widows had been married to the "single finest human ever created," which, while poignant, is actually an observed phenomenon known as husband sanctification, when a wife speaks of nothing but the best traits of her dead spouse. The term was coined by sociologist Helena Znaniecki Lopata, who conducted the other major study on widows in Chicago in 1971. Lopata suggested that husband sanctification served two purposes: it elevated the widow's own self-worth, since she was married to such an exemplary man, and it turned the husband into a benevolent spirit so that the past could be remembered in an exclusively positive light. Like Silverman, Lopata concluded that the death of a husband led to a massive identity crisis, a prolonged state from which it was almost impossible to recover, but she also allowed that this might be more social convention than psychological reality. "Widows are expected to be devastated, even years after the death of their husbands," she noted. "Some respondents expressed concern about undertaking certain activities, lest they be criticized by neighbors or other members of the community for moving out of mourning too soon." Lopata's questionnaire also opened the door for widows to feel anything other than sorrow or desolation. Half of her respondents agreed with the statement "I like living alone," and only one fifth of the 290 widows answered yes to the question: "Would you like to remarry?" (Lopata's survey excluded women who had remarried.)

And here lies one of many paradoxes of the conventional wisdom about widows: if women are so dependent on their role as wife, to the point where their sense of self is imperiled by the death of their spouse, then why is it that a large portion of them don't get remarried and thus restore their identities? Why do male widowers get remarried more frequently and more quickly than female widows? One explanation is since women have a longer life span, and men often marry younger women, there are simply fewer potential new husbands on the market for widows. But upon closer inspection, there are other plausible reasons. Lopata and Silverman both noticed that some widows enjoyed their new independence, especially being able to do things and go places that their husbands wouldn't have liked. They also noted that older widows feared becoming the caretaker for an aging man if they got remarried. Adult children may be opposed to remarriage, and there are financial disincentives, such as losing your dead husband's Social Security payments. ("I'm never getting remarried, because I don't want to lose that check," Sue, one of the Widowbago widows in Fort Lauderdale, said. "It's the *only* benefit of widowhood.")

The conflicts about remarrying after widowhood might also stem from our tendency to idealize the institution as the manifestation of true love. In 1980, Robert Di Giulio was a consultant and author of parenting books with a Ph.D. in human development and education when his wife, his eldest daughter, and his in-laws were killed in a car accident. His focus turned to studying loss, which resulted in his book *Beyond Widowhood: From Bereavement to Emergence and Hope*. Di Giulio interviewed eighty-three widowed women and was surprised to discover that one half of them, while still saddened, also expressed relief following the death of their husband, predominantly from "doing what someone else wants," "cooking food I don't like," "taking care of a sick man," "doing all the dirty work," or "raising the kids with no help from him." These sentiments were expressed by women who nonetheless characterized their marriages as happy. "Widowhood was an opportunity for self-expression and self-discovery," Di Giulio wrote.

According to Di Giulio, a widow has to answer a critical question: Am I to be a widow for the rest of my life or not? To answer yes is to run the risk of shutting oneself off from further self-discovery. Di Giulio stressed that widowed people and all who come into contact with them must be taught that widowhood is only a temporary transition, not a permanent condition. Unfortunately, friends and family often encourage a woman to remain in the widow role. "This may be motivated by a wish to prevent her from violating the singular-love fantasy our society holds so dearly: 'you can truly love only one in this life,' " he noted. Di Giulio himself remarried four years after his wife's death and before his book was published, and went on to have two more children, although he doesn't mention this perhaps for the very reason that he didn't want to alienate readers who still cling to the one-soulmate-for-life concept.

Finally, there's a more subtle way that our grief culture has made widowhood particularly punitive. According to the prevailing theory, you don't want to rush grief because the pain of the experience is actually a good thing. Not only is pain good, but as the advice books repeatedly remind you, it is the *only* way out of grief. It is a message that some widows, understandably, find discouraging. "I hated all the books, hated them," says Maggie Nelson Burchill, whose husband, Trevor Nelson, a producer for *60 Minutes,* died in 2003 after being treated for viral meningitis (Burchill pursued a malpractice suit against two hospital employees). "I desperately did not want to be in this place of deep despair forever. The only thing that was helpful was a woman who wrote me a letter saying that she had been engaged to someone who had died, and while the loss was terrible, she wanted me to know that she was now married to someone else and doing okay. That was something I could hold on to. I was only thirty-six. I wanted hope."

3

The Work of Grief

In the 1950s, before Jessica Mitford's critique of the funeral industry led to congressional hearings, a sociologist by the name of LeRoy Bowman interviewed bereaved families to find out what kind of sales pressure they encountered as they made burial arrangements. "In the negotiations between buyer and undertaker, the latter stresses the lasting comfort that comes from the last look at the face of the deceased," he noted. According to the language of the day, an open casket supposedly gave friends and relatives a "memory image," but as Bowman pointed out, the claim that a glimpse at a dead person's face could provide lasting consolation was largely a ploy to try to tack on additional services and fees.

"The alleged solace for the bereaved to come from this incident strengthens materially the funeral director's urging that the body be retained above ground for three or more days. There then ensues the need for embalming, restoring, and adorning the body with an expensive casket and accoutrements, as well as for all the succeeding stages of a 'fine funeral.' "

Since those days, Americans have been steadily rejecting such costly services in favor of "direct disposition"—the cremation rate has risen from 3.6 percent in 1960 to 35 percent in 2008 and is projected to reach 55 percent by 2025. Funeral homes have responded to this growing threat with updated admonitions borrowed from the grief culture. "Many people, grief experts included, feel the ability to view the body gives friends and loved ones a necessary opportunity to say goodbye," claims a flyer intended for people considering cremation published by the National Funeral Directors Association (the NFDA). At the 2009 annual convention of the organization, which counts over ten thousand funeral homes as members, two such grief experts, the husband-and-wife team of Susan Zonnebelt-Smeenge and Robert DeVries, were being put into service to conduct a seminar on how to talk customers into holding a viewing even if they had already decided against a burial. "Despite those seeking to minimize the value of funeral events by selecting direct cremation, funeral events involving viewing and ritual are critical to helping a family begin a healthy grief journey," was how the seminar was billed. "Discover five reasons why funeral events are important for grieving loved ones and how you can communicate those reasons to a family seeking direct cremation."

The seminar wasn't the only evidence of the commercialization of grief at the convention. In the Boston Convention Center's Expo Pavilion, where vendors displayed tools of the trade such as burial vaults and embalming fluid, there were also several booths selling grief materials, including one outfit called the Grief Store. "We're the only ones who offer books, DVD and CD, something for every media style," said Darcie Sims, the founder and co-owner who is also the president of a consulting business called Grief, Inc. As I was talking to Sims about what funeral homes stand to gain by educating their customers about grief—"I say it's the best advertising that they can buy" said Sims—I noticed the potent odor of baked shortening that floats through an airplane's cabin at the end of a long flight when the attendants heat up chocolate-chip cookies for the first-class passengers: Otis Spunkmeyer had a booth across from the Grief Store and, with one of its signature mini-ovens in operation, was handing out samples. I crossed the aisle to find out what precut cookie dough had to do with the mortuary arts and found a promotional brochure offering the following explanation: "Funerals and wakes bring families together. Fresh-baked cookies offer the same warmth and nostalgia the bereaved seek during this trying time."

The partnering of grief with the funeral industry is a relatively recent phenomenon. According to Jessica Koth, the public relations director for the NFDA, grief products made their way into funeral home offerings about fifteen years ago. "People want one-stop shopping, so they look to funeral directors for that kind of help," explained Koth. In what is known in the industry as "aftercare services," funeral homes routinely send customers cards on anniversaries of the death, host Christmas memorial services, and even offer group and individual grief counseling. Many now have designated bereavement coordinators on their payroll and one, Carmon Funeral Home in Windsor, Connecticut, even co-founded a grieving center called Mary's Place. The relationship goes deeper than mere convenience: both the grief culture and the funeral industry hinge on the premise that the decisions that you make after the death of a loved one, beginning with how to handle the arrangements for the body, can positively or negatively impact your psyche.

Perhaps one of the reasons grief specialists don't seem to have qualms about being used to sell funeral home services is that they too have a vested interest in reaching potential customers at the earliest possible moment. "We first started coming to the NFDA convention in 1999," Susan Zonnebelt-Smeenge told me when I stopped by her booth, for in addition to giving the seminar to funeral home employees on how to talk people out of choosing direct cremation, she and her husband, Robert DeVries, were also selling their four books and other "helpful resources" (workshops, training videos, etc.) at the Expo Pavilion. (Zonnebelt-Smeenge and DeVries first met through a grief group after they were both widowed and became friends and writing partners before getting married.) "Grief is something that you have to work on," DeVries said. "The analogy that we use is if you broke your leg you wouldn't just sit there and do nothing." Their first book, *Getting to the Other Side of Grief: Overcoming the Loss of a Spouse*, comes with a set of assignments, such as "journaling," rereading sympathy cards, listening to a recording of the memorial service (they call it "the funeral tape") at least monthly, and writing letters to your deceased spouse. "The prescription to tackle grief work may seem strange to you," they say. "But if you're going to grieve anyway, why not work as hard as possible to help yourself?"

In fact, researchers have been casting doubt on the benefits of "grief work" for more than a decade. According to the grief work hypothesis, one of the worst things you can do is to try to resist your grief, a notion that originated with one of Freud's disciples, Helene Deutsch, who argued in 1937 that "unmanifested grief will be found expressed to the full in some way or other." But one study of thirty widows and widowers conducted by the husband-and-wife research team Wolfgang and Margaret Stroebe of Utrecht University in the Netherlands in 1991 found that widows who avoided confronting their loss were not any more depressed than widows who "worked through" their grief. One of the key components of grief work is to give voice to the loss, but several other studies done by the Stroebes indicated that talking about the death of a marital partner or otherwise disclosing emotions did not help people adjust to that loss any better.

Nonetheless, the belief in grief work seems almost as intractable as the belief that grief is a series of stages. The first laid the conceptual groundwork for the second, for if grief was work that, left uncompleted, could cause harm, it follows that people would need guidance in, and find reassurance from, knowing what steps they should take to avoid that eventuality. As psychologist Therese Rando warns in *How to Go On Living When Someone You Love Dies,* "[Grief work] demands much more than merely passively experiencing your reactions to loss: you must actively do things and undertake a specific course of thought and action to integrate and resolve your grief." (Along with Alan Wolfelt, Rando is one of the most frequently consulted grief specialists, so it should give people pause that they both continue to advocate grief work with such conviction.)

What that specific course might be depends on which of the competing schools of thought you consult. If you go the classic route, with Kübler-Ross's *On Grief and Grieving,* you will of course be instructed on the five stages, but you will also find the confusing caveat that they don't always happen in the same order, even though the word "stage" implies a specific sequence, start to finish. "We had a discussion about whether to change the term," says Kübler-Ross's co-author, David Kessler. "She had said once that 'stage' was just the word used in the day. I said that the word 'response' might be more accurate, but we were both of the mind that the stages were so ingrained in the culture, so prevalent in our society, that there was no pulling them back in." And so instead of abandoning the misleading term, Kübler-Ross and Kessler sprinkled it liberally throughout *On Grief and Grieving,* along with reminders of the importance of hitting each of them. "[Denial] is nature's way of letting in only as much as we can handle"; "Anger is a necessary stage of the healing process. . . . [It] means you are progressing"; "[D]epression is a way for nature to keep us protected by shutting down the nervous system so that we can adapt to something we feel we cannot handle."

If you'd prefer a newer, more updated approach, there are other options, but ever since Kübler-Ross introduced her five, the number of stages has grown substantially. Therese Rando proposed the six Rs: (1) recognize the death; (2) react (emotionally); (3) recollect and reexperience; (4) relinquish; (5) readjust; and (6) reinvest, which actually makes seven Rs by my count. In his book *Facing Death,* a former priest-turned-counselor named Robert E. Kavanaugh outlined seven phases: shock, disorganization, volatile emotions, guilt, loss and loneliness, relief, and reestablishment. In 2003, a widow named Annie Estlund, who wrote a book and started the Web site ForWidowsOnly.com ("Only widows really understand widows"), increased the number of stages to ten—shock/numbness, confusion, denial, bargaining, anxiety, anger, guilt, depression, cockiness, acceptance—which makes for a pretty exhausting itinerary. Like Kübler-Ross's original five, these additional stages were all based on anecdotes and personal experience, not methodologically sound surveys.

Undoubtedly, some people do experience some of these stages. The problem is not so much the emotions they describe, but that the overall structure implies a lockstep progression for a condition that, it turns out, is much less orderly. It helps to illustrate this point. If one were to diagram the stages of grief—for simplicity's sake, I'll stick with Kübler-Ross's five—the emotional trajectory over a period of several months would look something like this:

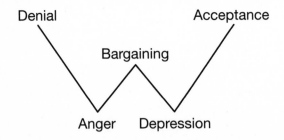

But the experience actually looks more like this:

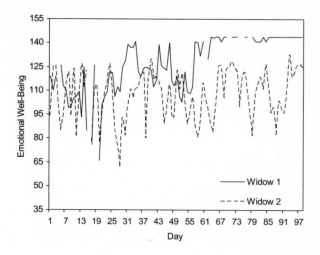

Emotional well-being of two widows, in which a higher score indicates better well-being. Gaps in Widow 1's trajectory are due to missing data.

We know this because a psychologist named Toni Bisconti, now at the University of Akron in Ohio, realized that while researchers had looked at grief both cross-sectionally and longitudinally, no one had actually measured it frequently enough to know how it unfolded in real time. So she and her colleagues asked recent widows to fill out questionnaires on their moods every single day for three months, and there were vast fluctuations. A widow might feel anxious and blue one day, only to feel lighthearted and cheerful the next.

(Bisconti noted that the fluctuations might even happen several times a day, but her study was not designed to capture that.) The swings started out quite large and then gradually diminished over time, getting both weaker and less frequent. Bisconti concluded that her results most closely resembled a linear oscillator, or a pendulum with friction, that looks something like this:

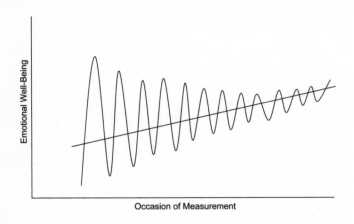

Occasion of Measurement

"Grief is anything but linear, and my data showing consistent ups and downs are obviously in conflict with stage theory," Bisconti told me. "Stage theories are also conducive to self-fulfilling prophecies and confirmation biases. In other words, if I lose my partner/spouse and I am angry on a given day, I'll think I'm in the anger stage and discount the fact that also on that day I might be sad, distraught, even happy at a given moment. We're probably all of those at different points." In 2008, psychologist Dale Lund of California State

University surveyed 292 recently bereaved men and women over the age of fifty and found that 75 percent of respondents reported finding humor and laughter in their daily lives, and at levels much higher than they had expected. But stage theory prioritizes negative emotions over any positive ones that might occur, say, from a happy memory of the deceased.

This oversight is particularly worrisome because it seems to be the positive emotions and not the negative ones that are the most healing. George Bonanno has found that not only is it normal for widows and widowers to smile and laugh when describing their relationship to the deceased, but that those who were able to do so six months after the loss were happier and healthier fourteen months out than those who could only speak of the departed with sadness, fear, or anger. As Bonanno wrote in his book *The Other Side of Sadness,* C. S. Lewis was on to something when he described grief as a "bomber circling round and dropping its bombs each time the circle brings it overhead." Bonanno added, "It is that respite from the trench of sadness that makes grief bearable. It is the marvelous human capacity to squeeze in brief moments of happiness and joy that allows us to see that we may once again begin moving forward."

So why does stage theory remain so entrenched? Partly, it's because our minds like to bring order to grief's contradictions. But as Toni Bisconti points out, "Stage theories are great for people who happen to experience emotion consistent with them and incredibly pathologizing for those who don't." Stage theory also fulfills a broader psychological need, the "human desire to make sense of how the mind processes, and may come to accept, events and circumstances that it finds wholly unacceptable," as researchers Holly Prigerson and Paul Maciejewski put it in the *British Journal of Psychiatry* in 2008.

Indeed, the stages of grief have been such a hit that they've jumped the track to all sorts of transitional life events, such as divorce, corporate burnout, even going away to college for the first time. (That's another reason we love stages so much—they can be applied to virtually anything.) In its orientation literature, Loyola University in Maryland informs incoming freshmen that they may go through stages similar to Elisabeth Kübler-Ross's. "At first you may feel shock and denial, then anger, then bargaining ('I'll give it another week and then I'm leaving'), depression, and finally acceptance." Stage theory has become so embedded in the self-help movement that we readily accept that in order to reach a particular goal—corporate success, financial freedom, you name it—we have to take a series of steps, usually numbering four, five, or seven.

But stageism isn't merely a mainstay of pop psychology—it has also made damaging inroads into the behavioral sciences. Consider the story of how Vivienne Cass, an Australian psychologist, developed her prominent model of coming out of

the closet: Cass was working as a therapist for the Homosexual Counseling Service of Western Australia in the 1970s
when a young woman came into her office full of self-loathing
because she was attracted to another female. Cass, herself a
lesbian, was curious about why some people accepted their
homosexual identity while others were so tormented by it,
so she started taking copious notes on how her clients spoke
about their growing awareness of their sexual orientation.
One day, as Cass would later describe, "I realized that certain phrases were being expressed repeatedly, simple phrases
such as 'I don't want to be different' and 'I'm only a lesbian
because of this particular woman.' After writing down these
phrases, and staring at them long and hard, I suddenly realized that there was actually a pattern to the way in which they
were expressed. Or, to be more precise, not so much a pattern as an order. Some of the cognitive insights I was hearing,
it seemed to me, only ever appeared in people's speech after
other particular thoughts were expressed. . . . I took to listening and observing further, and although I tried to find alternative patterns of thought processing in my clients, I kept
coming back to my original ordering. This simple list of
phrases of speech then became the basis on which I developed
my model of the six stages of homosexual identity formation."

Cass recalled that after she published her theory in the
Journal of Homosexuality in 1979, she received plaudits from
counselors all across North America and Western Europe

telling her how much the model "helped them to understand their client, and expressing amazement that a theory developed on the other side of the world could so closely fit their own experiences." In other words, counselors liked the theory because it helped the counselors. It made their work easier because it allowed them to place their clients within Cass's framework and it confirmed and clarified their own interpretations.

The Cass Theory, as it is now called, begins with Identity Confusion (asking oneself, "Am I gay?"), which is then followed by Identity Comparison, or looking to gay and straight peers in an effort to answer that question. Next comes Identity Tolerance, or seeking the company of other homosexuals but remaining uncertain about whether such an orientation is good or bad. Then Identity Acceptance, followed by Identity Pride, which is characterized by gay activism. People either work through each stage, remain stuck at a particular stage and stay partially in the closet, or undergo something called "identity foreclosure," which basically means that they slam the door shut. If they make it to the sixth stage they will reach Identity Synthesis, when homosexuality is an important aspect of the self but not a person's whole identity.

You can see why the Cass Theory became so popular: it turned a complicated experience into an orderly and predictable progression with a beginning, middle, and end. And it came with a built-in reward for completing the stages, since those who reach Identity Synthesis were supposed to be better adjusted—more peaceful and stable, less angry or afraid—as well as more integrated into society.

Two decades later, three researchers from the University of North Carolina at Greensboro decided to revisit the Cass Theory, noting that despite Cass's own limited evidence supporting her six stages, it had become the classic outline of homosexual identity formation and remained the most widely used model. The Greensboro group wondered in particular whether the decrease in the social stigma of homosexuality over the intervening twenty years had made some of the stages irrelevant. Also, since Cass's research pool was largely gay men, they wanted to see how her theory stacked up against the female experience. They conducted in-depth interviews with twelve women who identified themselves as gay to see which stages they recalled going through and found no consistent pattern whatsoever. Two women who were quite young when they first realized that they felt affection toward other girls (six and eight years old) skipped stages 2 and 3— Identity Comparison and Identity Tolerance—completely and went right to stage 4, Identity Acceptance, which was explained by their having already incorporated a knowledge of being different from their peers very early on. For most of the women, stages 3 and 4 were reversed, in that they generally did not seek out other lesbians until they had accepted their own lesbianism. Finally, most of the women never saw their sexual orientation as their primary source of identity or became activists, a hallmark of stage 5, Identity Pride.

The majority skipped that stage and went directly to stage 6, Identity Synthesis. The Greensboro researchers theorized that stage 5 might simply be a historical artifact from the gay rights movement of the 1970s that was no longer necessary as gay men and women became more accepted by society. Nine of the twelve women had reached stage 6, Identity Synthesis, but all twelve said that concerns about hate crimes and personal safety prevented them from complete public integration of their homosexual identities, showing that new inhibitions have also come into play since Cass came up with her theory.

In sum, then, only two of the stages—stage 1 (Identity Confusion) and stage 4 (Identity Acceptance)—were experienced by every woman in the study, which is merely stating the obvious: First people are uncertain or conflicted about being gay, and then they're not. This isn't to say that there are not some lesbians to whom the Cass model might apply, but the Greensboro study makes it clear it's hardly universal, which raises the question: If a model can't be relied upon to predict behavior for at least the majority, what purpose does it serve? Once again, psychological reassurance. "Theories provide a sense that others have passed this way before and made it, and that what one is experiencing in a given stage is 'normal,' " the Greensboro researchers concluded in 2000. "Cass's theory of sexual identity development offers that to some degree. . . . However, it seems to be time to update this theory or at least expand on it."

In 2009, I contacted Suzanne Degges-White, who led the

study while a doctoral student at Greensboro and is now a professor at Purdue University, to ask if the Cass model had, in fact, been updated as she and her co-authors had recommended. "I know that many researchers have found enough variations in identity development that new models are hypothesized in the literature, but few seem to have the name recognition that the Cass model has," she replied, before adding some personal insight: "I myself have to laugh at the whole notion of 'stages,' especially as it relates to my own coming-out process. I was in my early thirties and a married mother of three when I revealed to my mother that I could no longer suppress my true sexual identity. Her reply, of course, was 'Suzy, it's just a stage.' Well, of course every period in life is a 'stage' in some way!" When I asked Degges-White why stage theories in general are so popular, she offered this: "I wonder if we seek out stage theories as a way to make sense of changes that we don't fully control. And I wonder if when we say that *people may cycle through the stages* at different rates, or in different orders, or skip a stage, or whatever, that we are really trying to just fit a one-size-fits-all theory to a population for whom one size does *not* fit all. It seems that for humans—or at least those of us here in the Western world— if we can create a series of stages to account for an event that disrupts the normal flow of life, then perhaps this allows us to feel as if we are getting back some control."

*　　*　　*

This impulse originated in the 1930s, when Jean Piaget, the Swiss psychologist, invented a stage theory for cognitive development in children. Piaget was trying to keep his baby, Jacqueline, entertained in her bassinet one day when he observed what he would later describe as the first signs of real intelligence. "Jacqueline tries to grab a cigarette case I present to her," he wrote in his notebook. (Today we'd more likely use a soft, nontoxic toy devoid of unhealthy associations to distract a fussy child, but this was 1925.) "I then slide it between the crossed strings which attach her dolls to the hood. She tries to reach it directly. Not succeeding . . . she looks in front of her, grasps the strings, pulls and shakes them, etc. The cigarette case then falls and she grasps it."

Piaget saw young Jacqueline, then eight months old, demonstrating an awareness of cause and effect: If I pull the strings on my bassinet, the case will fall. But in Piaget's view, this accomplishment wasn't just an amusing milestone. As the first sign of an infant's capacity for logic, as well as a sense of agency in the world, it was an important building block for future cognitive development. Reaching this skill in what Piaget would later dub in his characteristically dry fashion "the fourth sub-stage of the sensorimotor stage" (typically between the ages of nine and twelve months) was an essential precursor to more abstract and sophisticated thought. Piaget subsequently constructed an all-encompassing theory out of such stages that essentially launched the entire field of developmental psychology. His theory, which consisted of four major stages and multiple substages, also set the ground rules for future stage theories: they are hierarchical, in that later stages grow out of earlier ones, and they are intransitive, that is, unable to be reordered.

Jacqueline wasn't the only child of Piaget's to become part of his living laboratory. Her younger brother and sister, Laurent and Lucienne, were also observed by their father (and were given equally hazardous objects, including matchboxes and penknives, to manipulate). In fact, for a while Piaget's only research participants were his own three children, a sample that was hampered not only by its small size but also its lack of genetic or environmental variation. His theory was eventually faulted as too tidy, and he subsequently admitted that, in truth, there were often unpredictable gaps in the sequence—he called these gaps "decalages"—which made his stages more approximate than absolute. Other developmental theories followed—Erik Erikson's eight stages of psychosocial development, Lawrence Kohlberg's six stages of moral development—but the further stage theory got from a child's physical growth, the more it lost its inherent logic.

This trend in academic psychology would eventually find its way to grief through the work of John Bowlby, the British psychiatrist who famously invented attachment theory after he observed how young children responded when removed from their mothers and placed with strangers. In 1960, Bowlby made waves by arguing that the "processes of mourning" in adults was akin to the three phases of separation anxiety that he had delineated in young children: pro-test, despair, and detachment. What's more, the way a child persistently seeks to reunite with his mother (there was little mention of the father's role)—and the resulting "unfavour-able personality development"—was also typical of the sort of mourning that goes awry in an adult. Bowlby's application of child psychology to adult grief attracted the attention of psychiatrist Colin Murray Parkes (who worked at the first hospice in the Western world, St. Christopher's in London) and who joined Bowlby's research unit at Tavistock Clinic. Together, they interviewed twenty-two widows, and came up with four phases of grief: numbness, searching and yearn-ing, depression, and reorganization. You will, on occasion, find Bowlby and Parkes's phases cited instead of Elisabeth Kübler-Ross's stages—on the Web site of the National Can-cer Institute, for example, which is part of the National Insti-tutes of Health.

But Kübler-Ross had a lead on Bowlby and Parkes out of the gate: she had already published *On Death and Dying* by the time Bowlby and Parkes introduced their four phases of grief in 1970. Kübler-Ross's stages were also more precise and memorable than Bowlby and Parkes's phases ("acceptance" just sounds better than "reorganization"), so they were easier to explain to the public. And so even though her stages were intended to describe how people face their own death, it was her theory, not Parkes and Bowlby's, that came to dominate our view of grief for the next half century.

4

The Making of a Bestseller

"There's this lady in Chicago, man...wrote a book...
Dr. Kübler-Ross with a dash. This chick, man, without the
benefit of dying herself, has broken the process of death into
five stages: denial, anger, bargaining, depression, acceptance.
Sounds like a Jewish law firm."
  —Lenny Bruce character in *All That Jazz* (1979)

"There are five stages to grief. Which are (*reads from computer
screen*) denial, anger, bargaining, depression, and acceptance.
And right now, out there, they are all denying the fact that
they're sad. And...that's hard. And that's making them all
angry. And it is my job to get them all the way through to acceptance. And if not acceptance, then just depression. If I can
get them depressed, then I will have done my job."
  —Michael Scott (Steve Carell) in *The Office* (2006)

The town of Seneca Falls in upstate New York is best known
for two things: hosting the first Women's Rights Convention
in the United States in 1848, organized by Lucretia Mott and
Elizabeth Cady Stanton, and being the inspiration for the fictional town of Bedford Falls in *It's a Wonderful Life*. It is not
easy to get to, and I was surprised to learn that a Japanese

doctor named Ryoko Dozono had traveled here all the way from Tokyo to attend an induction ceremony for Elisabeth Kübler-Ross into the National Women's Hall of Fame in October 2007. Kübler-Ross had herself passed away three years earlier at the age of seventy-eight, so Dr. Dozono, who runs a holistic health center in Japan, wasn't even going to get a chance to see her personally. "I met her on May 16, 1995," she later told me, remembering the precise date when, after several failed attempts at getting an audience, she was finally received into Kübler-Ross's home in Phoenix, Arizona. A friendship developed, and Dozono made several more visits to Phoenix and attended Kübler-Ross's funeral in 2004. The posthumous induction of her personal hero into the pantheon of prominent American women could not be missed. "I flew to Seneca Falls from Tokyo just to attend the ceremony," she explained. "It was my mission."

Dr. Dozono wasn't the only Kübler-Ross pilgrim at the National Women's Hall of Fame. There was Brookes Cowan, a social worker and sociology lecturer at the University of Vermont, who had met Kübler-Ross while filming a documentary in 2003 and received a call from Kübler-Ross's son, Kenneth, the following year to come and coordinate his mother's hospice care during the last week of her life. There was a playwright from Washington, D.C., who'd woken at four o'clock that morning to drive the eight hours to Seneca Falls because she'd been unable to find an empty hotel room in the area for the night before. And despite the fact that such prominent figures as Julia Child and Henrietta Szold, the founder of the women's Zionist organization Hadassah, were also being honored that day (both also posthumously), it was clear that Kübler-Ross was the star of the event. When Swanee Hunt, the former ambassador to Austria and director of Harvard's Women and Public Policy Program, mounted the podium to receive her own induction, she said, "When I was in my twenties, there was one person who could put chaos into order and that person was Elisabeth Kübler-Ross. I don't have words to tell you what it means to be named in the same paragraph as Elisabeth."

Kübler-Ross fans often talk about her in religious terms. Kate Eastman, who attended the induction luncheon and runs a home for dying children called the Jason Program in Portland, Maine, had a mystical experience after reading *On Death and Dying* as a Bates College student. "The voice was very specific: 'You will start a pediatric hospice in Maine,'" Eastman says. "I remember thinking: Huh? What? *Me?* I believed the voice to be that of God and that this was what I was called to do." Brookes Cowan saw Kübler-Ross speak at the University of Virginia in the late 1970s and "after that lecture, I embarked on a career in the field of end-of-life care, one of a million points of light ignited by Elisabeth," she wrote in a tribute to her hero. After the ceremony, I spoke to Dr. Ryoko Dozono by phone and when, at the end of our conversation, I thanked her for her time, she replied, "We have to thank Elisabeth—she made this opportunity for us to talk together."

I suppose it's not surprising that Kübler-Ross drew people to her via God's voice, people who see a million points of light and continue to refer to her in the present tense years after she died. Kübler-Ross transformed herself from a pioneering psychiatrist who exposed the insensitivity of her fellow doctors into a New Age healer who attended séances, sought guidance from two spirits named Salem and Pedro, and in 1977 declared that death did not exist. Her reputation suffered greatly as a result—one of her former research assistants published a paper arguing she was not really a doctor but a charismatic religious leader—but it has since been partially resurrected. In 1999, *Time* magazine proclaimed her an "unsung hero" in a special issue devoted to "The Great Minds of the Century." Her nineteenth and final book, *On Grief and Grieving,* published in 2005, got respectable reviews and sold about 65,000 copies. At the Women's Hall of Fame ceremony, there was no mention of her extended detour into the paranormal. Instead, she was remembered and honored for her famous seminars at the University of Chicago in which she interviewed dying patients; her resulting first book, *On Death and Dying,* which was published in 1969 and sold millions of copies; and her most lasting legacy: the five stages of dying and how they are now associated with any major loss.

To her supporters—and there are many—remembering Kübler-Ross only for her stages is a big mistake. In the fortieth anniversary edition of *On Death and Dying* (which costs $140) Allan Kellehear, a sociologist, argued in the Foreword that "The fundamental value of this work lies in the dialogue between two people discussing the meaning of dying. . . . To view Kübler-Ross's conceptual contribution as solely a contribution to theories of grief and bereavement is to dislocate and remove her work from the context of early hospice and palliative care research into dying. This is literally a confusion that some workers in the death, dying and bereavement fields constantly make in relation to this work." When I reached Kellehear by e-mail, he told me that although he'd never met Kübler-Ross, he had tried to speak on her behalf "to give her a bit of a voice if she were around to defend herself. Most people who criticize her have never spoken to even one dying person and don't understand the subtleties and complexities of this kind of work."

It is certainly true that people interpreted Kübler-Ross's stages more literally than she may have intended. But she herself played an active role in blurring the line between facing one's own death and facing the death of a loved one. In 1981, during an extensive *Playboy* interview when she could have set the world straight that her stages were for the terminally ill, she instead encouraged their misapplication, saying, "Even though I called it the stages of dying, it is really a natural adjustment to loss. Some people go through it if they only lose their contact lens." At the time she was defending herself against a larger public relations problem: her affiliation with a con artist named Jay Barham who claimed he could channel the dead. In that same interview, she said, "I have always been skeptical, a superskeptic," but then goes on to argue that negative thinking causes cancer and that cigarettes cannot kill you if you are not afraid of them. Kübler-Ross could have done more rigorous research to support or clarify her stages, but she didn't. In fact, during her entire career, she never published a single study on death or grief in a peer-reviewed journal.

Despite her medical training, Kübler-Ross was always more of a spiritualist than a scientist, more a believer than an empiricist. "My experiences have taught me that there are no accidents in life. The things that happened to me *had* to happen," she wrote in her autobiography. But Kübler-Ross's famous five stages and their widespread acceptance were, in fact, largely accidental. And although her stages would later be taken as absolute fact, the story of their genesis shows just how theoretical they really were.

People who knew Kübler-Ross often describe her as an inspiring public speaker who could bring her audiences to tears, but was stubborn and combative in person. In her autobiography, she painted herself as an iconoclast born into a conservative Swiss family who had to struggle to establish an identity separate from her two triplet sisters, Erika and Eva. By the sixth grade, Kübler-Ross recalled, she knew that she wanted to be a doctor, but not because she'd had any positive experiences with any. On the contrary, an early memory of being hospitalized for pneumonia was highly unpleasant. "I was weighed, poked, prodded, asked to cough and treated like a thing rather than a little girl as they sought the cause of my problems," she later wrote. Only her dying roommate, just two years older, provided comfort. "She said that her real family was 'on the other side' and assured me that there was no need to worry. I had no fear of the journey my new friend was embarking on. It seemed as natural as the sun going down every night and the moon taking its place. The next morning I noticed that my friend's bed was empty. None of the doctors or nurses said a word about her departure, but I smiled inside, knowing that she had confided in me before leaving. Maybe I knew more than they did."

Kübler-Ross peppered her autobiography with anecdotes like these in which nameless oracles appear—dying little girls, Polish Holocaust survivors, hospital orderlies—and share their wisdom with her. She dreamed of becoming a country doctor, but when she fell in love with a classmate of hers at the University of Zurich medical school, an American named Manny Ross, her plans got derailed. The Rosses, as they were then known, moved to Long Island where they both had internships at Glen Cove Community Hospital. She was accepted for a residency in pediatrics at Columbia-Presbyterian but then discovered that she was pregnant, which the hospital didn't allow because the schedule for pediatric residents was considered too physically demanding. Desperate for a job, she took a residency at the psychiatric ward of Manhattan State

Hospital. Then she had a miscarriage. Kübler-Ross would eventually give birth to two children—a son, Kenneth, and a daughter, Barbara—but she would also suffer a total of four miscarriages, an experience, she later said, that made her accept death as part of life's natural cycle. "It was the risk one assumed when giving birth as well as the risk one accepted simply for being alive."

As a psychiatry resident, Kübler-Ross was interested neither in psychoanalysis nor in drug therapy, the two main avenues of treatment at the time. In her work at Manhattan State Hospital, then a warehouse for the seriously disturbed, her approach seemed to consist mostly of talking casually with her patients and battling with her boss about trying to reduce their medication. In 1962, she and her husband moved to Colorado to take positions at the University of Denver Medical School, where she found a kindred spirit in an unconventional professor named Sydney Margolin, who ran a lab devoted to the study of psychosomatic medicine. Margolin asked Kübler-Ross to take over one of his lectures while he traveled on business and told her to pick any subject she wanted. "For the next week, I planted myself in the library and plowed through book after book, trying to find an original topic," she later recalled. "My experiences had taught me that most doctors were far too detached in their approach to patients. They desperately needed to confront the simple, down-to-earth feelings, fears and defenses that people had when they entered the hospital. So what, I asked myself, was the ground everyone had in common? No matter how much literature I looked at, nothing came to mind. Then one day the subject popped into my mind. Death." She decided to bring a sixteen-year-old girl with terminal leukemia whom she had met on her rounds into the lecture hall, with the goal that the young doctors-in-training would have no choice but to put themselves in the dying girl's shoes. Afterward, Kübler-Ross reported, the students sat "in a stunned, emotional, almost reverential silence. They wanted to talk but didn't know what to say until I started the discussion. Most admitted that Linda [the girl with leukemia] had moved them to tears. Finally, I suggested that their reactions, while instigated by the dying girl, were in fact due to an admission of their own fragile mortality. 'Now you are reacting like human beings instead of scientists,' I offered."

Professor Margolin returned from vacation, but Kübler-Ross had discovered a lecture strategy that, three years later, would attract Loudon Wainwright Jr., the editor of *Life* magazine, to write a story that brought her worldwide fame. "From an inconspicuous existence in which I did my own thing . . . I found myself featured in headlines all over the world, and the recipient of a flood of mail that has not ceased since then," she wrote in 1980.

But before all that happened, she and Manny moved to Illinois where she joined the psychiatric department of Billings Hospital, associated with the University of Chicago. Meanwhile, a transcript of her interview with the dying girl back in Denver made its way into the classrooms at Chicago Theological Seminary. That fall, in 1965, four male students came to her office asking if she would help them talk to terminally ill patients for a research project on the spiritual needs of the dying. "They were my students," remembers Phil Anderson, a former professor at the seminary who now lives in a retirement home for religious professionals called Pilgrim Place in Claremont, California. "I'd heard Kübler-Ross was over at Billings and I said, Why don't you go see her?" She found them a dying patient, and soon word spread and she had more requests from students. Eventually Kübler-Ross was leading a seminar every Friday, conducted in a teaching auditorium with a smaller room outfitted with one-way glass and an audio system where she would interview one dying patient while her audience looked on. The hospital's chaplain, the Reverend Renford Gaines, joined her for the interviews. Afterward, they would escort the patient back to his or her room and then return to the lecture hall for a discussion session with the students. "They were very interesting and really showed for the first time that dying people had many thoughts and feelings, something the medical establishment wasn't interested in at the time," says Phil Anderson, who attended many of the seminars. "She was a very feisty woman; she said that when she first asked the other doctors for people to interview, they told her they had no dying patients."

In December 1966, Kübler-Ross wrote an article for the *Chicago Theological Seminary Register* about these seminars.

Even though the article was called "The Dying Patient as Teacher: An Experiment and an Experience," it reads more as a manifesto against modern medicine, picking up on the idea of death as taboo laid out by British sociologist Geoffrey Gorer in 1955. "Dying is still a distasteful but inevitable happening which is rarely spoken about," she wrote. "One might think that the scientific man of the twentieth century would have learned to deal with this uniform fear as successfully as he has been able to add years to his life-span, or to replace human organs, or to produce children through artificial insemination. Yet, when we compare dying in less civilized and less sophisticated countries, we cannot help but see that we, in the so-called advanced civilization, die less easily. Advancement of science has not contributed to but rather detracted from man's ability to accept death with dignity." Kübler-Ross suggested that her fellow doctors were unable to help their patients die because they saw death as a failure and were projecting their own anxiety on the experience.

In the article, Kübler-Ross described her interviewing technique. She approached potential subjects by saying she was looking to speak to "very sick or terminally-ill patients" and never used the words "dying" or "death" unless the patient did. The interviews started with some practical questions such as, How long have you been in the hospital? How do you deal with your illness? Occasionally a student would sit in along with hospital chaplain Reverend Gaines. "This enabled the patient to sit and listen passively and to join in only when he felt ready to accept an interpretation or answer a question raised by one of the interviewers," she noted. None of the interviews lasted longer than forty-five minutes and there was no follow-up.

It's well known in the research community that this kind of unstructured conversation is open to interviewer bias, as compared, for example, to a standardized questionnaire where everyone is asked the exact same thing; it may yield some interesting case studies, but no valid data from which broad conclusions can be drawn. And yet these interviews were the basis for the stage theory that Kübler-Ross subsequently proposed in *On Death and Dying*. As Allan Kellehear commented in his introduction to the fortieth anniversary edition, "It is essential to note that . . . *On Death & Dying* is not a work of research. It is a popular book of description, observation and reflection based on a series of dialogues with dying people. The participants were not invited to be part of a research project but were instead asked to talk about their experience to assist health professionals to understand their needs better."

Perhaps even more telling is the fact that by her own account, at the time of the actual interviews, Kübler-Ross was more interested in the reactions of the students and staff than any process the dying patient might be going through. Were the students afraid of the patients? Why were some of the nurses defensive when asked to help? By contrast, she observed, the patients almost uniformly accepted their own deaths: "They welcome a breakthrough of their defenses," she wrote in her *Chicago Theological Seminary Register* article. "They welcome a frank, unemotional, honest discussion and a sharing of their feelings." If you were to apply her own theory to the patients, you would say that they had all reached the final, fifth stage, except that the five stages are never mentioned, because, in fact, she hadn't conceived of them yet.

Meanwhile, back in New York, Peter Nevraumont, then a young editorial assistant at Macmillan Book Company, came across Kübler-Ross's article. Nevraumont was working in the philosophy and religion department where part of his job was to comb through academic journals for ideas. "We had discussed a book on death, so it was something we were looking for," remembers Nevraumont, who now runs his own small publishing house. "I read the article and it seemed like a book in miniature, so I wrote her a letter and said, 'Do you want to do a book?' and she said yes immediately."

Kübler-Ross was paid $7,000 for fifty thousand words, which she had promised she could deliver in three months. In her autobiography, she described struggling with formulating a concept. "It took three weeks of sitting at my desk late at night, while Kenneth and Barbara slept, before I figured out the book. Then I saw very clearly how all of my dying patients, in fact everyone who suffers a loss, went through similar stages. They started off with shock and denial, rage and anger, and then grief and pain. Later they bargained with God. They got depressed, asking, 'Why me?' And finally they withdrew into themselves for a bit, separating themselves from others while hopefully reaching a stage of peace and acceptance." In other words, the stages of death were the result of late-night brainstorming to overcome writer's block, rather than a consistent pattern derived from explicit interviews with dying patients.

The way Kübler-Ross described it, the stages came to her suddenly, almost as if through divine inspiration. But there is disagreement about their original authorship, with some people crediting an obscure psychologist named A. Beatrix Cobb of Texas Tech University. Known as BZ, Cobb had worked with cancer patients at the MD Anderson Cancer Center in Houston and developed a theory about stages of adjusting to one's own death that followed a progression from denial to anger to bargaining to acceptance—four stages in all. (Cobb was more a teacher and clinician than a researcher, so her stages were just as theoretical as Kübler-Ross's.) When *On Death and Dying* came out (with the additional stage of depression coming before acceptance), "people were encouraging BZ to take Kübler-Ross to task, but BZ said no," remembers Frank Lewis, a former student and colleague of Cobb's. "She just said that it made little difference to her who got the credit as long as people were benefiting from the knowledge or the work." Cobb died in 1990 at the age of eighty-three and never published any writing on stages, so there's no definitive proof that the idea was originally hers. But another former graduate student of Cobb's, Frank Lawlis, who went on to teach

TV personality and self-help book author Phil McGraw and is now the chief psychological advisor to the *Dr. Phil* show, remembers Cobb teaching her four stages as they applied to patients learning that they had terminal cancer in a class that he took in 1966. "I remember her example as she started her talk: 'What's the first word people are going to say when they receive bad news? No. That's denial.'" Lawlis received his Ph.D. in 1968, a year before the publication of *On Death and Dying*. "[BZ's theory] predated Kübler-Ross for sure," Lawlis told me when I reached him by phone. "But BZ didn't particularly want to say that stuff of hers was stolen because she was a very unselfish person." Lawlis went on to explain that Cobb spoke of the stages in existential terms and not as a literal progression. It was a big mistake, he said, that Kübler-Ross's version got taken at face value. "What sells is simplicity, making life a little more simple, so if you can give something that's very complex and individual and unique a simple plan, it'll stick."

There is no mention of BZ Cobb in the introduction or bibliography of *On Death and Dying,* and no way of knowing whether Kübler-Ross had heard of Cobb's stages or was completely ignorant of their close similarity to her own. It's certainly not unheard of in academia for the same idea to percolate up from different sources. But British psychiatrist Colin Murray Parkes also contends that Kübler-Ross owes him a big debt. Although Parkes and John Bowlby didn't publish their theory on the four phases of grief until 1970,

Parkes says that he was introduced to Kübler-Ross during a visit to the University of Chicago in May 1965. "While there I gave a lecture about my work with Bowlby and she subsequently adapted our 'phases of grief' and applied them to the approach to death in her book *On Death and Dying,*" Parkes wrote to me in an e-mail (Bowlby died in 1990). "She was not very good at acknowledging sources of help and had a somewhat abrasive relationship with her medical colleagues." Parkes and Bowlby's phases were based on interviews with only twenty-two widows, not a particularly large or diverse sample on which to base a universal theory, but which also means, if Parkes's account is correct, that the stages actually did a loop-de-loop from grief to death then back to grief again.

Kübler-Ross did thank Peter Nevraumont, the young editorial assistant at Macmillan, for suggesting that she write the book. "I guess I had a good instinct in those days and it became a huge bestseller, not that it accrued to me in the least since they were paying me $110 a week," Nevraumont told me. "But it didn't surprise me that it did well. It just seemed like demographically it was coming at the right time. The baby boomers' parents were starting to die. Today it's the baby boomers watching their mates die and trying to deal with that."

The article on Kübler-Ross appeared in *Life* on November 21, 1969, creating a perfect storm of publicity coinciding with the publication of *On Death and Dying*. Very soon afterward, Kübler-Ross's interests took an un-academic turn into life after death. She decided that her next project would be to find and interview people who had been revived after their vital signs had flatlined, and enlisted Reverend Gaines as her collaborator. But that project never got off the ground, because by early 1970, Reverend Gaines left Billings Hospital to take over a church in Urbana. (He later adopted the African name Mwalimu Imara.) Gaines's boss, a pastor named Carl Nighswonger, took his place in the death and dying teaching seminars, but he and Kübler-Ross did not get along. "There was such a lack of chemistry between us one student mistakenly thought Pastor N. was the doctor and I was the spiritual counselor," Kübler-Ross recalled in her autobiography. "It was dismal." The enmity was apparently mutual. According to John Rea Thomas, then president of the Association of Mental Health Chaplains, Nighswonger complained that not only did he and his associate chaplain at the hospital, Herman Cook, do all the legwork finding the dying patients for the seminar, but that the patients were often upset afterward and that Nighswonger and Cook "would have to go back and put the pieces together after Kübler-Ross was done with them." Thomas also remembers that both Nighswonger and Cook felt that Kübler-Ross had appropriated the stages of dying from BZ Cobb.

In later interviews, Kübler-Ross said that she was ousted from the University of Chicago because she hadn't published enough academic research. (She also said that she had already decided to leave but that a woman named Mrs. Schwarz whom she'd interviewed for the seminar and who had passed away came back from the dead to tell her "You cannot stop this work on death and dying, not yet.") But according to C. Knight Aldrich, who as chairman of the psychology department at the University of Chicago had originally hired Kübler-Ross, the real problem was that her demand on the lecture circuit was causing her to neglect her teaching duties. "She was assigned to a seminar for first-year students that observed expecting mothers, but she didn't show up and I called her to account for that," Aldrich told me. "She said, 'But that's just eight students; I was talking to eight hundred nurses in Scranton about death and dying!' So when she came up for tenure, I said that I couldn't support her promotion because she was spending too much time away." Aldrich's reservations, while valid, also mirror a certain resentment of Kübler-Ross in academia, which seems to stem from her celebrity. One professor told me, "There's a saying on our field that Kübler-Ross wanted disciples, not colleagues. Well, she got them."

After Kübler-Ross departed Billings and the university, Carl Nighswonger took over the weekly seminars with dying patients, which was by then known as the "D&D" class. By 1971, Nighswonger had replaced Kübler-Ross's five stages with his own theory that he called the six dramas of death, adding the stage of "celebration" after acceptance for those patients who seemed able to celebrate life in their last days.

He also specifically brought family members into the equation. "The important thing, in Nighswonger's opinion, is that families begin their grieving before the patient dies and do it openly," wrote David Dempsey in an article profiling Nighswonger in *The New York Times Magazine* in 1971. According to Nighswonger, soon-to-be-bereaved family members should synchronize their own emotions with those of their loved one in order to *help* them die. "This is done by encouraging a 'natural' response—uncontrolled weeping, anger, silent grieving or simply a heart-to-heart talk." Nighswonger, who was in his early forties, seemed poised to give Kübler-Ross a run for her money, but on May 13, 1972, he died suddenly of a heart attack, a few days before a scheduled appointment with a cardiologist.

Kübler-Ross quickly followed up with two sequels to her bestseller, *Questions and Answers on Death and Dying* and *Death: The Final Stage of Growth*. She also began to lecture about guiding both the dying and their loved ones through the five stages, according to an account of a two-day conference titled "The Patient, Death and the Family" held at the University of Rochester Medical School featuring Kübler-Ross as the keynote speaker. In 1974, Richard Schulz published his article raising concern about the lack of evidence for the stages, but it was too late: they had already reached critical mass and skeptics were left trying to make sense of their popularity. Mark W. Novak, a sociologist now at the University of Winnipeg, theorized that the stages functioned as myth that facilitated a discussion about a difficult topic, with Kübler-Ross as the master storyteller. "By schooling the clinician in the 'stages of dying,' Kübler-Ross helps free us from our fear, anxiety, and guilt in the presence of the dying," he wrote in the *Psychoanalytic Review* in 1979. "Yet a danger lurks here—the danger that these stages themselves will be taken . . . as a scientized series of steps that define or help diagnose the extent of a person's development toward death."

In fact, that's exactly what did happen, although not as the result of any one particular event but as the theory gained wider and wider circulation. In 1969, Dr. Marjorie C. Meehan opened her review of *On Death and Dying* in *JAMA* by writing, "Most people, even doctors, don't like to think, talk, or read about death. Even so, I hope a great many will read this book." Kübler-Ross's extensive speaking and writing about the stages stimulated a kind of applied thanatology that then spread to every single discipline of the helping professions—social work, psychology, pastoral care. "Kübler-Ross is the one person that anyone who has read anything on dying and death has read," says George Dickinson, a professor of psychology at the University of Charleston who has done extensive surveys of death education over the past two decades. By the time the theory reached high schoolers such as myself, it had been watered down to become, simply, the stages of grief.

As for Kübler-Ross herself, she became a firm believer in life after death and opened a "healing and growth" center called Shanti Nilaya in California in 1978. In a lecture in San

Diego, she announced "my real job is . . . to tell people that death does not exist," a conclusion she had arrived at having heard as much from "our friends who have passed over, people who came back to share with us." A year later, readers of *Ladies' Home Journal* nominated her a "Woman of the Decade," but soon a scandal erupted when it was discovered that one of her favorite healers had been pretending to channel the dead husbands of female visitors as a ruse to get them into bed with him. Kübler-Ross had to close her center, and eventually moved to Virginia and opened another one called Healing Waters, where she continued to give workshops on "life, loss and transition." In October 1994, her house and center burned to the ground, and all of her papers, including twenty-five journals in which she recorded all the conversations she'd had with her spirit friends Salem and Pedro, were destroyed. In 1995, Kübler-Ross suffered a major stroke and was partially paralyzed, but after moving to Arizona to be closer to her son, Kenneth, she kept writing books and posted personal news for her followers on her Web site. After 9/11, she visited Ground Zero in a wheelchair accompanied by state troopers and clergymen and met with recovery workers at St. Paul's Chapel. In 2004, after proclaiming for years that she was ready to expire, Kübler-Ross died at her home in Scottsdale at the age of seventy-eight.

5

The Grief Counseling Industry

"I said in the first edition of this book 25 years ago that I don't believe that we need to establish a new profession of grief counselors. I still believe this."
—J. William Worden, in the introduction to the 2009 edition
of his book *Grief Counseling and Grief Therapy: A Handbook
for the Mental Health Practitioner,* the manual for
the bereavement field

At a professional development course held in the Dallas Hyatt Regency, a group of aspiring and practicing grief counselors sat debating what to do about tissues. If clients start crying, do you hand them a box or will that interrupt their emotions? "I've seen instances where when you hand people a tissue and they're really in the flow of something that it just stops them," said the session's instructor, Valerie Molaison, a psychologist and the clinical director of a bereavement center for children and families in Delaware called Supporting Kidds.

One student raised her hand: "What if you just put the box of tissues next to them but don't hand it to them directly?"

"If you're offering them tissues, then you're telling them to stop crying, that it's too messy," responded Regina Franklin-Basye, a hospital bereavement care coordinator taking the class. Finally, a solution was reached: Don't move the tissue box next to the client but keep it visible and within their grasp at all times.

What wasn't being debated, but probably should have been, is whether counseling people through their grief actually works. Grief support of one kind or another is now routinely offered in a range of settings, beginning with the places where most people die: hospitals, palliative care units, and, most of all, hospices, where it's mandated for a minimum of one year after a loved one's death ever since Congress passed Medicare hospice legislation in 1982. As of 2008, approximately five thousand paid full-time hospice employees are devoted to bereavement services, which usually include phone calls, visits, and mailings to family members, and that's not even counting the estimated 550,000 hospice volunteers who also interact with grieving families. (Hospices are also mandated to use volunteers for 5 percent of all patient care hours.)

If family members don't encounter someone offering to

help them in their grief at the place where their loved one died, they surely will at one of the nearly 21,000 funeral homes across the country. "It's safe to say that most reputable funeral homes offer some kind of aftercare program, whether it's personal correspondence through phone calls or written communication, or having an aftercare coordinator/grief counselor on staff to facilitate bereavement," says Emilee High, the communications director of the National Funeral Directors Association. (In its guidelines for "aftercare," the industry-approved term for bereavement services, the NFDA does caution funeral home employees to not misrepresent themselves as counselors unless they're licensed, and steers them toward other descriptors—"Funeral homes may use the label 'grief facilitator' to describe the funeral director's role.")

Religious professionals of all stripes minister to people in grief as a routine part of pastoral care, especially chaplains working at hospitals and retirement homes. Support groups are frequently hosted at churches and community centers, not to mention online, where there are dozens of Web sites devoted to all sorts of loss. Finally, there are freestanding organizations across the country—from the Greenwich Village Center for Separation and Loss, to the Center for Grief Recovery in Chicago, to the New Hope Grief Recovery and Support Community in Orange County, California—that offer a wide array of treatments. (Some are not-for-profit, but others are all for it: the Grief Recovery Institute in Sherman Oaks,

California, has trademarked the term "grief recovery" and charges $995 for a two-day workshop.) Because grief counseling draws from so many different disciplines (nursing, social work, psychology, and religion), it's impossible to precisely tabulate how many practitioners are out there. When I asked several veterans in the field for their best guess, they came in at anywhere from 20,000 to 100,000 people who specialize, or at least devote the majority of their working lives, to bereavement support.

For a practice that has become so ubiquitous, it has been awfully hard to verify its effectiveness. Years of outcome research shows that traditional psychotherapy, where people seek help for a variety of conditions, has been proven to work; regardless of the type of therapy or whether that therapy targets an individual, family, or group, studies usually conclude that treated clients are better off than their untreated counterparts. By contrast, grief counseling is intended to help just one problem, and as such, it's considered something of a preventive measure: in order for it to be successful in a quantifiable way, it should either speed up normal grief or help people avoid complications such as long-term clinical depression or intense, prolonged suffering. But in 2008, after Robert Neimeyer, a psychology professor at the University of Memphis, and his colleague Joseph Currier decided to analyze the results of over sixty controlled studies on grief intervention, they found no consistent pattern of an overall preventive effect. "What that means is, instead of finding that people who received counseling got better or stayed the same, and that people who didn't receive counseling got worse or stayed the same, we found that everyone just got better," Currier says. One-to-one counseling wasn't the only intervention that they examined. Their survey included a wide range of treatments: guided imagery, cognitive behavioral counseling, peer groups, psychological debriefing, even supportive phone calls. On average, those who got help experienced no less distress nor recovered more quickly than those who didn't. Their study concluded, "Such evidence challenges the common assumption in bereavement care that routine intervention should be provided on a universal basis or according to special circumstances surrounding the loss"—for example, all parents who lose a child.

The only instance where counseling was measurably successful in the aggregate was when it was targeted at people who were having prolonged difficulties. Although they weren't always explicitly assessed for it, this group was probably showing what researchers have come to call complicated grief. This is defined as an acute state that impedes functioning for longer than six months, and is characterized by intense yearning for the deceased and distressing and intrusive thoughts about his or her death. Since complicated grief affects only approximately 10 to 15 percent of the grieving population, it is considered an abnormal response in the sense that it deviates from the statistical norm. "The conclusion is that in this point in time, given the current research, we cannot say that grief counseling is as effective with adults who are showing a normative response," Currier says. (Currier and several colleagues published a similar analysis of interventions for children, and although the pool of existing controlled outcome studies was much smaller, it yielded a similar conclusion that they weren't, on average, beneficial.)

It would seem pretty crucial information that the only people grief counseling may help are the people who are the worst off, but there was no mention of these findings in the professional development course that I attended. But that's hardly surprising either, as it took place at the 2009 annual conference of the Association for Death Education and Counseling (ADEC), the closest thing grief counselors have to a trade organization. Since its founding in 1976, ADEC has grown to become the central nervous system of the grief culture and publishes two journals, *Omega: The Journal of Death and Dying* and *Death Studies,* whose contents make for some pretty bleak reading with occasional high absurdity, such as one article, "The Association Between Thoughts of Defecation and Thoughts of Death." (In case you are wondering, there seems to be an inverse correlation: more death, less shit; more shit, less death. "It is proposed that this tendency for keeping thoughts of death and feces separate is part of a broader inclination for keeping the sacred from the profane," the author concluded.)

Currier had presented his analysis of outcome research on grief interventions at the previous ADEC conference, to a decidedly mixed reception. "Some people were excited, but I think some people dismissed it because it didn't confirm what they were doing," he says. In fact, Currier's findings present an existential crisis for the grief counseling industry in general, and ADEC in particular. In addition to holding yearly conferences, ADEC is also in the business of minting grief specialists with a certificate in thanatology, or awarding them a CT. In order to take the CT examination, which is administered once a year online, a candidate needs simply a bachelor's degree and two years of related experience (which can include, for example, volunteering at a hospice). You must also rack up a certain number of "contact hours" by paying to attend conferences and seminars such as the two-day professional development class I attended. Other than that, the requirements are fairly slim—no master's-level degree or clinical supervision required. ADEC certifies about 150 people each year, and although a CT is not a license to provide counseling—those are given out state by state and have varying requirements—it does enable them to call themselves bereavement educators and run support groups through hospitals, churches, or funeral homes, which essentially puts them on the front lines for encountering the bereaved. As ADEC says in its vision statement, the goal of the organization is to "provide a home for professionals from diverse backgrounds to advance the body of knowledge and to *promote practical applications* of research and theory." (Emphasis mine.) That means keeping the bar of entry as low as possible and not limiting credentials to people with advanced degrees in mental health.

Because of this loophole, when certified thanatologists or CTs come across a case of what may be complicated grief, they are supposed to refer the client to a trained therapist such as a clinical social worker, psychologist, or psychiatrist. (To her credit, my instructor, a licensed psychologist specializing in childhood bereavement, emphasized this point in our grief counseling class.) But if that 10 to 15 percent who are doing poorly are supposed to be referred to a trained mental health practitioner for therapy, and the remaining 85 to 90 percent who react normally don't get any earlier relief from counseling, then one might wonder what is the point of grief counselors at all. To this, our instructor said, "If someone's not pathological, are we still going to be helpful? We all know the answer to this is yes. Grief is like childbirth: you can go out in the woods and squat, but you can also go to the hospital and get help." When asked what he would recommend to a counselor in light of the results of his study, Joseph Currier, himself an occasional counselor, said, "You need to have different goals. If your goal is to make the person better when there's a strong likelihood that the person is going to get better on his or her own, then you have to be honest with yourself. Part of your work may be to help them see that they already have the resources that they need." For a counselor advising someone who is grieving normally, that would mean telling the client that they'll do just fine on their own without counseling.

But that wasn't the message being delivered in my grief counseling class. (If it had been, we could have adjourned much earlier and gone to the old book depository at Dealey Plaza to visit the museum that now occupies the sixth floor from which Lee Harvey Oswald shot JFK.) Instead, we spent the bulk of two days getting schooled in the Four Tasks of Mourning, invented in 1982 by J. William Worden, a psychologist in practice in Laguna Niguel, California. Although Kübler-Ross's stages of grief are better known to the public, Worden's tasks of grief lend themselves particularly well to counseling because they promote the notion that the mourner isn't just a passenger along for a ride, but can (and should) be an active participant while at the same time enlisting outside help. As Worden writes in his handbook for counselors, "The tasks concept is much more consonant with Freud's concept of grief work and implies that the mourner needs to take action and can do something. Also, this approach implies that mourning can be influenced by intervention from the outside." Or as my instructor said to us, "Tasks help you organize what can be a very chaotic experience for everyone. Keep these tasks in mind when you're seeing people, because if you can begin to create order from chaos, it helps them a lot."

The four tasks that mourners are supposed to tackle (and counselors are supposed to help them accomplish) are: (1) to accept the reality of the loss; (2) to process the pain of grief; (3) to adjust to a world without the deceased; and (4) to find an enduring connection with the deceased in the midst of embarking on a new life, although Worden has changed this last task significantly since he first conceived of his model. In 1982, the fourth task was "withdrawing emotional energy from the deceased and reinvesting it in another relationship." This was consistent with Freud's theory that the goal of grief work was to "decathect" from the dead, a notion that became passé in the 1990s when researchers discovered that people continue to think about and even carry on conversations (in their heads or out loud) with their deceased loved ones in a way that isn't necessarily delusional. So in 1991, Worden softened the fourth task to "emotionally relocate the Deceased and Move on with Life." The theory that people maintain bonds with the dead continued to get more popular, and in 2008, Worden changed his fourth task yet again to "Find an Enduring Connection with the Deceased in the Midst of Embarking on a New Life." The problem with the latest iteration is that it assumes that this is something that everyone should and would want to do, and that it's universally beneficial, when in fact encouraging someone to cultivate an ongoing connection may keep them in grief longer than is necessary. George Bonanno and colleagues assessed thirty-nine bereaved people and found that those with the strongest "continuing bonds" also had the most elevated grief symptoms up to five years after the loss, suggesting that an ongoing attachment to the dead could be, just as Freud originally thought, unhealthy. Nonetheless, continuing bonds are still very much in favor— "It's the opposite of what people used to think," our instructor noted—so in our class we were given some ideas for helping clients foster them, such as suggesting that they write letters to the deceased, or invite them to important events.

Forty years ago, there was no such thing as a grief counselor (or grief educator, companioner, facilitator, thanatologist, etc.). That role was usually played by someone in a person's existing support network, or perhaps a priest or psychotherapist. But as the study of death and dying exploded in the 1970s—Vanderlyn Pine, a former funeral home director turned sociology professor, called it the period of "pop death"—the field developed its corps of grief professionals, even while a great many of them were attracted to such work for extremely personal reasons. ADEC was founded in 1976, and its membership grew dramatically in the late 1980s and currently stands at more than 1,700 people, although there are also many counselors who are not members. In 1989, ADEC introduced its certification program. At first, master's degrees were required (along with letters of recommendation and relevant experience) in order to be awarded a CDE (Certified in Death Education), CGC (Certified in Grief Counseling), and CGT (Certified in Grief Therapy), but eventually they lowered those standards for their current CT program. "We decided [against a master's requirement] because there are numerous professionals in our field who have only the BA (nurses, funeral directors, BA social workers, hospice bereavement folks), and for whom certification would be valuable and a way to encourage them to improve their own knowledge about thanatology," says Gordon Thornton, a professor emeritus of psychology at Indiana University of Pennsylvania and former president of ADEC. "We decided *against* certification in any titles with 'counselor' or 'therapist' in them because we do not assess clinical competency." Nonetheless, people with CTs come into contact with grieving people all the time, and there's understandable confusion over how counseling is officially defined (a support group leader might call himself or herself an "educator," even though they are effectively a group counselor) and who is qualified to give it.

As the ranks of grief counselors grew, the media reinforced their practice as the new norm. In 2008, a sociology graduate student named Laurel Hilliker analyzed all the articles on grieving that were published in *The New York Times* between 1980 and 2006 and found that 66 percent mentioned or encouraged counseling of some kind. "Consistent throughout the reports were recommendations for outside help, which was seen as an important component in doing one's 'grief work,' whether through support groups or individual counseling," Hilliker noted. The majority of experts cited or quoted in those articles were, of course, grief counselors, although occasionally there were contrarian voices in the mix, such as Dr. Sally Satel, a psychiatrist who questioned the need for such a specialty. "Are our priests and rabbis not up to the task? Are our families' instincts to comfort not keen enough?" Satel asked in a 1999 column. "The deployment of counselors—a well-meaning effort, I wholly grant you—sends an odd message: that people are too fragile to soothe and strengthen themselves." Satel eventually co-authored a book called *One Nation Under Therapy,* in which she argued that our dependence on outside counselors, a trend she called "therapism," was eroding our self-reliance. But her argument didn't get to the crux of the matter, which was (and continues to be) that nobody had conclusively proven that counseling for grief is measurably beneficial.

Even William Worden, the inventor of the Four Tasks of Mourning and an advocate of counseling if ever there was one, doesn't believe in a specialized corps of grief counselors. In his introduction to *Grief Counseling and Grief Therapy: A Handbook for the Mental Health Practitioner,* published in 2009, he reiterated the position he took in the first edition of the book in 1982, which is that there was no need to establish a new professional group. There were plenty of people such as doctors, nurses, social workers, and clergy who were "already in a position to extend care to the bereaved and have the knowledge and skills required to do effective intervention and, in some cases, preventive mental health work." Worden's book was the only required reading for my counseling class, so having noticed his caveat in advance, I was curious how it would be received among my fellow classmates, about half of whom had no clinical or pastoral training at all, including myself. (Others were more qualified professionals such as psychiatric nurses and social workers. There were also two members of the New York State police force whose job it was to contact the families of officers who died in the line of duty.) But very few of my classmates had purchased the book, much less read it. Meanwhile, our instructor seemed to have reached a different interpretation of Worden's warning when she told us, "Worden says you *don't* have to be a mental health professional to do grief counseling." Just to be sure about his position, I contacted Worden and told him that our instructor had invoked his blessings on those of us untrained in mental health counseling, and he said, "Absolutely not. I meant the exact opposite. The training level is extremely important. The analogy I use in my classes is, I can carve a mean turkey but that doesn't mean I can do surgery."

Forty years ago, Worden, then a young investigator at Harvard who went on to do a child bereavement study, befriended Elisabeth Kübler-Ross at a panel. Through that connection, he was invited to conduct some of the first continuing education classes for doctors and nurses on grief at the University of Chicago. "I never intended to spawn a whole new professional group or paraprofessional group," he says. "Obviously, grief counselors are now abundant, and the problem is that a lot of these counselors don't have training in psychopathology and can get into a whole lot of trouble with a client who is having a hard time." Worden is particularly wary of people who become drawn to counseling after experiencing their own, difficult experiences with grief. "It's a problem," he told me. "What I've seen in training workshops are people who have undergone their own losses and want to be helpful to other people to work out their own grief."

People get drawn to professions for all sorts of personal reasons—a victim of a violent crime may decide to become a prosecutor, a child who overcomes cancer may grow up to become a doctor—but such instances are fairly rare. With grief counseling, however, it's extremely common to find people who made a major career change and turned grief into their profession after their own experience with a traumatic or untimely death. Within my class of a dozen people, there were at least four who fit that bill, and those were just the ones who volunteered the information in casual introductions at the beginning of the course. One woman, Joyce Lennon of Wilmington, Delaware, lost her husband in a plane crash and is now an "aftercare coordinator" at a funeral home. "I went to a psychiatrist and he said, 'Your husband died, get over it,'" Lennon told me on our lunch break. "This was back when grief was just swept under the rug." Another woman, Mina Gates, was a tech salesperson until her husband died of lung disease and she became a hospice volunteer, got her CT, and is now getting a master's in counseling. Regina Franklin-Basye had started a nonprofit outreach ministry for women who lost their mothers after she lost her own suddenly to bone cancer. Then, after suffering several late-term miscarriages, Franklin-Basye began working with a hospital to develop bereavement programs for perinatal loss.

After the class was over, I called up Joyce Lennon to hear more about how her husband's death inspired her to go into grief counseling. It had been his second marriage, and after his company plane lost one engine and nosedived into the side of a mountain, Lennon discovered that he had never updated his will. Everything he owned—including the house Lennon was living in—now belonged to his former wife or children. At the time, Lennon was working as a PR director for a hospital. "One day, a moving van pulled up and his sons informed me that I was moving out," Lennon recalled. "It was a very traumatic experience that definitely compounded my grief. I had a breakdown and tried to commit suicide and wound up in the hospital." Eighteen months later, Lennon enrolled in Wilmington University to get her Bachelor of Science degree in psychology. "For two years I was known on campus as 'The Grief Lady' because every paper, every project I did was on grief." After graduation, Lennon was hired by a funeral parlor. "I don't call myself a grief counselor though, I call myself a grief specialist or facilitator, because the word 'counseling' suggests that you have a problem," she says.

Having loss in one's background could certainly be an asset to a grief counselor, but it also inevitably colors one's interpretations and recommendations. Not only have many grief counselors experienced traumatic loss but so has almost every prominent grief expert out there. Childhood leukemia claimed the best friend of Alan Wolfelt, whom we met in the first chapter (and who subsequently attended workshops with Elisabeth Kübler-Ross). Therese Rando, author of *How to Go On Living When Someone You Love Dies,* was orphaned as a teenager: her father dropped dead with no warning while she was in high school and less than a year later her mother died from medical negligence after a successful open-heart surgery. Hope Edelman, author of the well-regarded *Motherless Daughters,* lost her own mother when she was seventeen. Ann Finkbeiner, a respected science journalist, wrote *After the Death of a Child* after her eighteen-year-old son was killed when the Amtrak train he was riding on his way back to college was struck by a freight car. Linda Goldman, the author of *Life and Loss: A Guide to Help Grieving Children* and several other grief books, gave birth to a stillborn daughter. J. Shep Jeffreys, a psychologist and the author of *Helping Grieving People When Tears Are Not Enough: A Handbook for Care Providers,* lost his young son Steven to cancer. (Five years later, in 1980, he attended a weeklong residential workshop with Kübler-Ross and "exploded with the stored grief material of Steven's illness and death and also old grief related to the sudden death of my mother when I was 16.") And the list goes on.

Almost every person out there who has written a book on grief has experienced the sudden, unexpected, and often violent death of a loved one, so that extraordinarily difficult circumstances have formed the filter through which we have come to understand loss in general. In 2010, I called Vanderlyn Pine, the sociologist who had warned, back in 1977, that the death and dying movement was attracting people for emotional reasons. I asked him how personal experience with loss would color a practitioner's perspective. "The problem is that when people enter the field with a broken heart because someone close to them has died, they feel that they have paid their penance and therefore already know all that there is to know," he said.

The lack of neutrality among grief professionals wouldn't necessarily be an insurmountable problem if it were routinely acknowledged and specifically warned against, the way any psychotherapist candidate is required to undergo therapy to become aware of his or her own biases and is trained in the importance of not projecting onto their clients. But as John R. Jordan, a psychologist who runs the Family Loss Project in Sherborn, Massachusetts, has pointed out, "Most grief counselors would acknowledge that they are drawn to the work in part because they seek to deal with loss in their life . . . in the training that my colleagues and I offer, we have come to understand the grief counselor's loss experiences as primarily a resource to be skillfully used in the therapeutic encounter, rather than antiseptically guarded against by the clinician."

Using personal experience or anecdote instead of research to guide treatment has been a big problem with applied thanatology all along. As Jordan explained in an article criticizing the disconnect between research and practice, "Mental health professionals in general, and grief counselors in particular, have shown a penchant for adopting theory on conviction alone, mostly because it 'feels right.' In my experience, it is rare for therapists to wonder about the potential harm done to clients when caregivers operate on beliefs that 'everyone knows' to be true." This may be why explanations of how any particular counseling approach was developed—Was it based on research of the needs of bereaved people? How exactly was it supposed to help?—are hard to come by.

"There are too many theories out there, and none of them are ever tested," says Chris Feudtner of the Penn Center for Bioethics. Feudtner is a pediatrician who has treated hundreds of dying children at the Children's Hospital of the University of Pennsylvania. "I've always worried when people say, 'We want to support these families,' because just because it sounds good doesn't mean it's helpful." As someone who, in addition to his medical degree, has a Ph.D. in the history and sociology of science, Feudtner knows that "there are many examples in medicine where the idea sounded good but the treatment turned out to not be helpful." Even so, Feudtner followed the prevailing wisdom and started bereavement support groups at his hospital until a lack of encouraging results led him to wonder, What's the evidence? So in 2004, he did his own review of grief intervention efficacy studies (his included some without control groups, unlike Currier and Neimeyer's, which was strictly limited to studies with controls). His conclusion? Other than treating major depression with medication, there was no evidence for recommending bereavement interventions. Feudtner offered some compel-

ling reasons why: the interventions were all over the map—
"treatments featured in published studies vary almost as
much as the authors who tested them," he noted. There was
also very little description of the actual interventions themselves, which resulted in a dearth of replication studies, the
only way treatments can reasonably be codified. Finally, he
suggested that "the bereavement care literature may be too
invested in and reliant on theoretical justifications of treatments."

Feudtner still runs support groups for bereaved families,
but says that they're not very heavily attended. The way he
looks at it, any form of generalized treatment for grief is
likely to miss its target. "When you're trying to treat someone, you're trying to mechanistically make them better," he
explains. "If it's something simple, like pneumococcus in-
fects your lungs, we can kill that germ, but with something
like grief, we don't know the mechanism. The other thing is
that people have resources like resilience and strength and
will just get better on their own, and it's very hard to show a
treatment effect if most people just get better anyway. Most
people spontaneously recover from six months to a year. So
you can get rid of all of those people and just hone it down
to a group that's not getting better, but you're still stuck with
the first problem, which is, what's the mechanism that's going
to work? For one thing, these people probably weren't necessarily healthy to begin with. Are they having trouble because
they're in a ruminative thought process? Or are they socially

isolated and not taking care of themselves? Did they have certain issues that were compensated for in the relationship? There are a number of different ways for people to take the death of a loved one. What we really need is to know where they're coming from."

6

The Grief Disease
and Resilience

In 1991, at the thirteenth annual conference of the Association for Death Education and Counseling, Therese Rando, the psychologist in private practice in Rhode Island who invented the six Rs of grief, gave a keynote address titled "The Increasing Prevalence of Complicated Mourning: The Onslaught Is Just Beginning." Rando asserted that Americans were on the brink of an epidemic with disastrous consequences: complicated mourning was rising dramatically, although she didn't say by how much or mention who had actually measured the increase. Our failure to prepare for the coming onslaught, she warned, would "place our society at greater risk for serious sequelae known to emanate from untreated complicated mourning." (She did provide a citation for the "serious sequelae"—her own forthcoming book about treating complicated grief.)

Rando went on to catalogue the causes of the onslaught, many of which seemed to have little to do with bereavement: wife-beating, abortion, child abuse, the rise in TV violence, "an increase in conflicted and dependent relationships in our society." She finished her critique of society's ills by pointing the finger at her fellow grief therapists: they too were contributing to the increase in complications by misdiagnosing them and hampering treatment (which was counter-logical; usually practitioners contribute to the rise of a disorder by *overdiagnosing* it). Rando offered this solution: a category for complicated grief needed to be added to the *Diagnostic and Statistical Manual of Mental Disorders* published by the American Psychiatric Association (known as the *DSM* and updated about every fifteen years) so that clinicians would learn how to properly recognize and treat the oncoming horde of sufferers.

Grief made its first appearance in the *DSM-III*, published in 1980, in an entry for "Uncomplicated Bereavement," which was defined as a "normal reaction to the death of a loved one" that occurs within the first two or three months of the loss. The entry was relegated to the back of the manual in a section for conditions that don't qualify as full-fledged mental disorders such as malingering or senility. These conditions are known as "V codes" (because their coding begins with a V, not to be confused with the other kind of V codes, the three-digit verification numbers on the back of your credit card), and get little clinical attention because their treatment is not reimbursable by insurance. As of this writing, grief is still a V code (V62.82 in *DSM-IV-TR*), although practitioners are given a means to fudge by diagnosing major depression if the symptoms are still present two months after the loss, reflecting the belief of some psychiatrists that any significant distress stemming from bereavement can be captured by the already existing diagnosis of depression, and doesn't warrant its own entry.

Despite her concern about complicated grief, Therese Rando did not follow through on the lengthy process of getting a new disorder added to the *DSM*. That heavy lifting eventually fell to a member of the second wave of grief researchers to emerge in the 1990s, two decades after the beginning of the death and dying movement. In 1995, Holly Prigerson, then a psychiatric epidemiologist with a specialty in geriatrics, was sitting in a meeting of a group researching mood disorder in later life at the University of Pittsburgh as they reviewed data from a study trying to treat bereavement-related depression. Antidepressant medication and therapy seemed to be helping the sadness and lethargy of the depression, but not the yearning and searching of the grief. "The assumption was that grief was normal, and depression was pathological, so the psychiatrists were not concerned. So I decided to test to see if grief symptoms were distinct from depression symptoms, and whether grief predicted future impairment over and above that of depression and anxiety," says Prigerson, now at the Dana-Farber Cancer Institute in Boston. "The answers were yes and yes." Those results, published in the *American Journal of Psychiatry* in 1995, began for Prigerson what has turned into a decades-long campaign to create a full-fledged entry for complicated grief in the next edition of the diagnostic manual, *DSM-5,* which is expected to be released in 2013. Prigerson and her growing number of supporters say their motivation is benevolent: identifying the small percentage of bereaved people with the most distressing and disabling symptoms is the first step toward being able to understand their condition and help them cope better with their loss. Up until now, only one treatment has been designed specifically for this population, and it involves both "revisiting" the death repeatedly in addition to practical advice on reengaging in the world. Because that treatment is so far-ranging, it is hard to know which aspect of it is actually helpful.

If there has been major dissent within the field of grief, it is over whether grief of any kind—complicated or not—should be conceptualized, and treated, as an illness. In his article "Mourning and Melancholia," published in 1917, Freud characterized grief as a normal reaction, but also noted that it becomes "pathological" when the survivor has mixed feelings about the deceased, leading to guilt and unconscious self-reproach for his or her death. In other words, what created problems (in typically Freudian inside-out logic) was not how much you loved the person who died, but how much you secretly disliked them. The pathology stemmed from the psychiatric makeup of the survivor, specifically whether he or she had "a disposition to obsessional neurosis."

For the next several decades, what was known as "pathological mourning" was described in the clinical literature as delayed, absent, or excessive grief stemming from ambivalent feelings. In 1944, Erich Lindemann, a psychiatrist at Mass General, picked up on more peculiar deviations when he observed that some grievers feverishly busied themselves with activities formerly carried out by the deceased, or even acquired symptoms of the same illness that had felled their loved one. Lindemann's article, "Symptomology and Management of Acute Grief," marked the first approach to grief as a medical problem that needed treatment, and there was a good reason why: his case studies included people under his care in the hospital's psychiatric wing who were already mentally ill when they happened to lose a relative. This may have contributed to his conclusion that "patients with obsessive personality make-up and with a history of former depressions are likely to develop an agitated depression."

In the 1960s, the death and dying movement shifted the discussion of problems away from both psychoanalytic and psychiatric interpretations to look at external forces that might prolong or intensify grief. The most apparent cause, about which there was much agreement, was society's denial of death, which, in trying to hurry people toward healing, created "unacknowledged" or "disenfranchised" grief. As an antidote, the "naturalness" of grief, no matter how lengthy, was universally promoted, and the outward expression of it encouraged. The primary message was, *There is no right or wrong way to grieve, there is only your way.* (Except when that way was avoidance.) Even the terminology changed: "pathological grief," which assumed that the problems originated with the mourner himself or herself, was jettisoned in favor of the more blame-free term "complicated grief," which carried the additional benefit of implying that what was complicated could be made uncomplicated. As practitioners began to speculate about the causes of complicated grief, they focused on the specific details surrounding the death itself, such as whether it was sudden or violent, or if the person died very young.

This approach gave rise to several stereotypes. For example, you have probably heard that the death of a child is the hardest loss that one can experience (and that parents who lose children have a higher risk for divorce). This certainly sounds true and makes intuitive sense, in that no parent expects to see his or her own offspring, for whom their love is almost limitless, die before they do. In many movies and novels, losing a child is certainly depicted as having the most dramatic, long-term repercussions. We have even projected this interpretation onto animals: when a baby gorilla named Claudio died in the Münster Zoo in Germany in 2008 and his mother, Gana, carried his corpse around for three days, captivated onlookers assumed that despair was driving her behavior. Photos of the Gana appeared in newspapers across the world, with headlines such as "A Mother's Grief: Heartbroken Gorilla Cradles Her Dead Baby." Finally, primatologists pointed out that in the wild, apes often hold on to dead children for long periods, not because they're experiencing terrible sorrow but because, lacking the ability to take the baby's pulse, they're protecting their genetic investment in case the baby is just temporarily comatose.

The theory that child loss is worse than all others gained currency during the death and dying movement thanks to a 1979 paper published by Catherine Sanders. The idea was relatively modern: During the previous period of expressive grief, at the turn of the nineteenth century, infant mortality rates were still quite high and the death of a child more commonplace. Sanders, at the time a recently minted Ph.D. from the University of South Florida, had entered the field late in life, having decided to go to graduate school after her seventeen-year-old son died in a waterskiing accident. In the aftermath of his death, her daughter ran away from home, and Sanders and her husband subsequently divorced—two very serious repercussions indeed. Sanders wanted to be able to compare the intensity of grief after the death of a child with the grief of losing a spouse and losing a parent, but no standard measuring tool existed, so she created one she called the Grief Experience Inventory (GEI), consisting of 124 true-false questions about predominantly negative symptoms such as despair, anger, loss of appetite, and death anxiety. She then made use of newspaper obituaries to recruit and administer the GEI to 102 newly bereaved people in Tampa, and found that parents who had lost children scored much higher and seemed to be suffering much more intensely than people who lost a spouse or a parent. That, however, was as far as her research went. She only measured them once, and on average the initial interview took place about two months after the time of death. "It can be safely said that those who experienced the death of a child revealed more intense grief reactions . . . than did those bereaved who had experienced the death of either a spouse or a parent," she concluded. "It appears that the death of a child results in loss of emotional control, which exposes those survivors to greater vulnerability of external influences. This fact, coupled with the intense somatic experiences noted, would explain in part the feelings of despair and alienation expressed by parents whose child had died."

Sanders left academia for private practice, opened a clearinghouse to distribute and sell the GEI, and developed a stage theory of her own called the five phases of grief (shock, awareness of loss, conservation/withdrawal, healing/the turning point, and renewal, to which she was considering adding a sixth phase of fulfillment before she died in 2002).

She never published a follow-up to her Tampa study, so we have no way of knowing whether her subjects who had lost a child remained the most distressed one year, two years, even five years out. But Stanley Murrell, a psychology professor at the University of Louisville, assessed 130 people five times every six months and published a study in 1990 that showed that the death of a spouse caused longer-lasting depression than the death of a child. Murrell suggested that spousal loss required a more intensive restructuring of one's life and was more "globally stressful." He also pointed out that as painful as losing a child is, one at least has a spouse to lean on. Murrell's study is out of date and certainly not the last word on comparing child loss with spousal loss. Subsequent studies suggest that the death of an adult son or daughter results in more intense or persistent grief than that of a young child, regardless of whether the loss was from illness or a sudden accident, perhaps because parents have grown so attached and have made such a lengthy investment in their progeny. But the conflicting findings illustrate the problem of categorizing certain losses as more difficult than others—there are simply too many other variables involved.

Like other clinicians of her time, Sanders was operating under the assumption that complicated grief was quite common, with one third of all major losses thought to result in complications that require professional help. Over the years, though, the estimated prevalence figure for complicated grief has dropped from that high of 33 percent to about 20 percent in 2001, and is now hovering at around 10 percent of cases of the most commonly experienced kind of loss, late-life widowhood from natural causes. The name too has gone through several permutations, reflecting various trends in mental health. During the late 1980s, when trauma became a popular cause for many syndromes (including multiple personality disorder), complicated grief was renamed traumatic grief and conceptually likened to posttraumatic stress disorder. In 1997, Holly Prigerson, who had moved from the University of Pittsburgh to Yale, convened a panel of researchers, including several PTSD specialists, to establish the criteria for diagnosing traumatic grief. The resulting definition, which was published in 1999 in the *British Journal of Psychiatry,* was distinctly influenced by traumatology in that it was broken down into two components: symptoms of separation distress (yearning, searching, loneliness) and traumatic distress (numbness, disbelief about the loss, anger, and a sense of futility about the future). But it soon became apparent that the problem with pairing complicated grief with trauma was that sudden, violent, and unexpected deaths didn't always result in pathological reactions (and pathological reactions sometimes arose from deaths from old age or lengthy illnesses). Even the most sudden, violent incident of mass death to occur in modern-day America—the nearly three thousand lives lost in the terrorist attacks on September 11, 2001—didn't result in complicated grief for friends and family members of the victims the majority of the time.

Much has been written about the impact of those events on mental health. As Arieh Shalev, an Israeli psychiatrist, pointed out in *9/11: Mental Health in the Wake of Terrorist Attacks,* it was because most of New York City remained structurally intact after 9/11 that such an intense focus was placed on the psychological needs of its residents (and to a lesser extent, of residents of Washington, D.C.). Grief suddenly became a matter of public health, for which there was no shortage of professionals willing to offer services. In the first days after the attack, therapists of all stripes streamed downtown to volunteer with other "first responders" such as firefighters and police officers. One estimate put the number of therapists who made their way to the offices of the Red Cross, frequently on foot, at nine thousand, which, even in therapist-rich New York City, seems rather high. As Erica Lowry, a social worker who oversaw the 9/11 health services for the Red Cross, subsequently described the scene, these disaster mental health (DMH) volunteers were immediately integrated into every aspect of the agency's 9/11 response. Charged with the task of administering "psychological first aid," they were among the few civilians permitted to enter the red zone below Canal Street, where streets were closed and fires still burned. They conducted weeks of door-to-door outreach in lower Manhattan and staffed 101 service sites and sixty sheltering facilities in the metropolitan area, while their international colleagues in more than sixty other countries met with family members of the deceased.

But not everyone wanted, or needed, immediate counseling. Alissa Torres, a 9/11 widow whose husband, Eddie, began working at Cantor Fitzgerald in the World Trade Center on September 10, 2001, remembers volunteer counselors descending on the makeshift family center at St. Vincent's Hospital, one of the places where people went to search for their lost relatives. "While we organized to try and find our loved ones, they got in our way because they were so dependent on us to feel useful," she recalled in her graphic memoir, *American Widow*. "Later on, at Pier 94, the official family center, it was worse. There we'd meet so many more grief workers." In the months following 9/11, Torres received numerous phone calls and home visits from the Red Cross Mental Health Unit, but, with a newborn baby to take care of and facing eviction, what she desperately needed was money. "I would have to tell my story over and over again," she told me in a phone conversation. "Meanwhile, the counselors kept leaving phone messages saying, 'How are you feeling today?'" Finally, Torres received her first "Family Assistance Gift" from the Red Cross.

Therapeutic concerns were often at odds with practical logistics. When the city's medical examiner's office needed to create a database of information about the missing victims in order to be able to make matches with what little human remains were being found, Shiya Ribowsky, the medical examiner's director who was in charge of that immense effort, recalled that at the family assistance center at Pier 94, many mental health professionals "campaigned strongly to have the VIP [Victim Information Profile] form filled out by the family on paper with the help of an interviewer who was making eye contact with them," instead of the interviewer entering the information directly into a computer. "Their theory had it that this would diminish the likelihood of a family member suffering through a cold, impersonal interview. . . . The problem was that we ended up with 25,000 VIP forms on paper, each at least seven pages long." It took 180 federal disaster relief specialists almost two months of working around the clock to then enter the information into a database so that it was searchable by computer—two months during which family members had to wait before they could get proof that their loved ones had perished. "I often wonder what is more hurtful to a family member: having an interviewer enter answers into a computer as they are given with perhaps a little less eye contact or waiting months so that the data could be used?" Ribowsky later wrote. "It seems to me that well-meaning mental health professionals can lose the big picture the same way as anyone else."

Longtime veterans of disaster relief are often cautious about bringing in therapists early on because they can interfere with more immediate needs. In the aftermath of 9/11, these included communicating with loved ones (cell phone services were jammed), transportation out of the city (or into the city for families of the deceased), shelter for the evacuated, and finally, money. Writing about 9/11, Simon Wessely, a British psychiatrist and epidemiologist, argued that meeting those concrete needs should have had top priority over psychological support. "I continue to have some skepticism about the immediate role of the mental health professional in the acute drama, other than his or her role as a good citizen," Wessely wrote. "I am also aware that such advice is hard to implement when the need to do something is overwhelming." In Wessely's opinion, the best psychological support comes from family and friends, not therapists. "Simply offering to help does not mean that help is needed, nor that a particular individual is qualified to give that help." The time for therapy was later on, and only for those who were most troubled. "Once the dust has settled, literally and figuratively, those with defined psychiatric disorders can access decent quality treatment," he concluded.

Karen Seeley, a clinical social worker and psychologist at Columbia University who surveyed thirty-five volunteer 9/11 therapists, found that the bereaved "were advised to discuss their experiences immediately in the name of mental health," as she wrote in her 2008 book, *Therapy After Terror*. This approach had its origins in two schools of thought: One was that people who have been through a disaster need to tell their stories within several days lest their emotions "seal up" through critical incident stress debriefing, a method that has since been discredited. The other held that a voice must be given to everyone impacted, not just the people who lost friends and relatives but also the alienated and frightened public eager to mourn as a way to feel connected to the event. This impulse was generous and well-intentioned, but it blurred the line between who was truly suffering and who was merely participating in a communal keening.

For those who did actually lose a loved one, the outsider's attempt to empathize was a mixed bag, validating feelings of loss for some and bringing discomfort to others. As Danielle Gardner, whose brother Douglas Gardner worked at Cantor Fitzgerald and died in the World Trade Center, wrote in a remarkable essay published in 2005, "I have learned about the whacked-out phenomenon I term trauma envy. I have learned that Americans, New Yorkers, people, seem to have a need to make this tragedy theirs, to feel close to this, the most significant event of our lifetime. People will compete with you. They'll say they were this close to the buildings, they watched this much TV, they couldn't sleep for so long, they were almost in the subway at that time—as if those experiences somehow equate with having a loved one evaporate in the cauldron of those hellish buildings."

The singularly horrible circumstances of the attacks presented many new challenges to public officials, agency workers, and therapists dedicated to helping 9/11 families. "Those who specialized in bereavement were uncertain how to assist persons who had no body to bury, whose losses could be counted in dozens, whose private losses were strangely public, or whose relatives were killed by members of a previously unheard of international terrorist organization," Seeley wrote. Mayor Rudolph Giuliani asked his health commissioner at the time, a psychiatrist named Neal L. Cohen, to bring in bereavement specialists to brief him on the handling of families as well as to help craft his public statements. "The mayor was concerned about his messaging, specifically how best to speak of the tragedy with healing words while respecting the victims' families' bereavement process," Cohen later recalled.

Michael Cohen (no relation to Neal Cohen) was one of the psychologists brought in to advise Giuliani. Cohen had been called upon before, by the Port Authority of New York and New Jersey, after the 1993 bombing at the World Trade Center killed six people and forced fifty thousand workers and visitors to evacuate. The Port Authority's goal was to get the buildings reopened as soon as possible with a minimal amount of public panic, so they enlisted Cohen to research individual and group reactions to the bombing and come up with an overall communications strategy. At the time, people were not so much concerned with ongoing terrorist attacks as they were about the structural integrity of the building complex itself, which housed about forty thousand tenants. Cohen wound up basing his strategy on a model proposed by psychiatrist Judith Herman in *Trauma and Recovery,* her 1992 book on sexual and domestic violence and abuse. "The fundamental stages of recovery are establishing safety, reconstructing the trauma story, and restoring connection to the community," Herman wrote in her introduction. Cohen recommended that the Port Authority stick to those tenets and the World Trade Center was able to open after about a month.

Eight years later, in 2001, Michael Cohen again turned to this model in his advice to city officials. By that point, Cohen's focus had shifted from trauma to child psychology, and he was consulting for the Board of Education. At midnight on September 11, he received a call from New York City schools chancellor Harold O. Levy. Levy had recommended Cohen to Mayor Giuliani, and Cohen had mere hours to prepare for a meeting with the mayor. Early the next morning, a car arrived in front of Cohen's apartment in Brooklyn and took him to the Police Training Academy on East 20th Street in Manhattan. There, with Governor George Pataki looking on, he gave Giuliani recommendations based on his 1993 adaptation of Judith Herman's stages of recovery: You have to be a trusted voice, you've got to reconstruct the trauma story, and you should get people together as much as possible. According to Cohen, Giuliani immediately incorporated these directives into every facet of his 9/11 response, from the phrasing of his comments about the city's safety to the planning of vigils in public parks and squares. "His performance those three months was absolutely extraordinary," recalls Cohen.

Today, Cohen has his own business conducting research on children for such clients as PBS Kids and the U.S. Department of Education. His offices are located in a high-ceilinged loft space on West Broadway in SoHo, where vintage posters and maps of New York City decorate the walls. Huge iMacs sit on every desk. When I went to interview him, I immediately noticed a copy of Elisabeth Kübler-Ross's *On Death and Dying* on his bookshelf, between James Gleick's *Chaos* and D. W. Winnicott. When I asked him whether Kübler-Ross was ever referred to in the subsequent briefings he had with Giuliani and members of his team, he said not in so many words because her theories hardly needed to be articulated. "Everyone knows about the stages and that grief is a process," he told me. (When I had first told Cohen I was researching a book about grief, he shared with me that his father had died suddenly when he was four and his mother had died when he was a teenager.)

In the beginning, Giuliani was advised to anticipate denial from the victims' families and to let them down slowly, so he waited almost two weeks, twelve days after the last survivor was rescued from Ground Zero, before he declared that it was unlikely any others would be found. "I believe that it is certainly time to say that the chances of finding anyone would now involve a miracle," he said at a briefing on September 24. "We knew, because of the severity of the fire, that there was probably no possibility of finding any more survivors, but some of the families weren't ready to accept that," Cohen told me. So the city continued to call the cleanup at Ground Zero a "rescue-and-recovery" mission out of sensitivity to the families. "From our perspective, we were doing the same thing at the site anyway, so it didn't hurt us to keep that possibility open. Why take that hope away from them?" Cohen said.

It was the first of many decisions that burnished Giuliani's reputation as a sensitive leader, despite the fact that only a year earlier he had displayed remarkable insensitivity when he used a press conference to inform his wife that they were separating. Once Giuliani began to close the window of hope for the 9/11 families, he also took steps to prevent them from getting stuck in limbo by expediting the processing of a death certificate. Then, in early October, Giuliani announced that he would be giving families their own urns of Ground Zero dust (which had been blessed by a chaplain at the site). Michael Cohen found a grief counselor who specialized in memorials to create a round mahogany box inscribed with the date 09-11-01 for the ceremony at the end of October. Families were also given access to a viewing platform of Ground Zero, constructed at the corner of Liberty and West Streets, with counselors on hand to accompany them. "Giving survivors space while letting them know that you were nearby for emotional support was a difficult balance for many counselors to strike," recalled administrators who organized a series of visits for families in New Jersey. "For the most part, families drew emotional support from one another."

After the first few weeks had passed, a massive clinical effort was orchestrated. At health commissioner Neal Cohen's encouragement, a team led by Dr. Sandro Galea at the Center for Urban Epidemiological Studies at New York Academy of Medicine began random-digit-dialing adults living south of 110th Street in Manhattan between five and eight weeks after the attacks to gauge the psychological impact of the event in order to plan services. They found that out of a sample of nearly one thousand people, approximately 10 percent had a friend or relative who had died in the attacks, and of those, 11 percent reported symptoms consistent with PTSD, and 18 percent had symptoms consistent with major depression. However, 7 percent of people who did *not* lose a friend or relative in the attacks also screened positive for PTSD, and 9 percent screened positive for depression, figures that were still higher than the non–post-9/11 rates Galea used as benchmarks, 3.6 percent for PTSD and 4.9 percent for depression in the past thirty days. (The lifetime rate for depression is much higher.) "How long the psychological sequelae of the September 11 attacks will last remains to be seen, and it is possible that the prevalence of symptoms in our study reflects transient stress reactions to some degree," Galea and his co-authors concluded, before pointing out that the ongoing threat of more attacks as well as the lengthy clean-up and disruption of services might prolong or worsen symptoms. "In this context, the high prevalence of psychopathology that we documented among the residents of Manhattan is not surprising."

So what happened? The New York State Office of Mental Health launched Project Liberty—its motto was "Feel

Free to Feel Better" and featured public service announcements from Susan Sarandon—in November 2001. With $155 million from the Federal Emergency Management Agency (FEMA), it became the most expensive counseling program in history. Anyone within the entire metropolitan region, which included many counties in Westchester and the Hudson Valley and New Jersey, was urged to call an 800 number to get a referral for free counseling from one of the hundreds of agencies contracted by Project Liberty. (New York City's health department coordinated a similar publicity campaign with the tagline "New York Needs Us Strong," a message that was intended to destigmatize accepting help.) Project Liberty prepared for 2.5 million potential clients, their estimate for the number of those directly impacted by 9/11.

But soon it became clear that there were more services than people making use of them. Project Liberty's director, April Naturale, began sending counselors door-to-door in downtown Manhattan and stationed them on busy thoroughfares throughout the rest of the city. And it was not the 9/11 families who were seeing counselors the most but other groups: People who lost family members accounted for 40 percent of Project Liberty counseling sessions in the first month, but that figure dropped to 5 percent or fewer by five months. (Displaced employed and unemployed workers, people with disabilities, and uniformed personnel used the services the most.) When the Red Cross joined forces with the September 11th Fund in August 2002 to cover mental health treatment costs for family members of the deceased, evacuees, and rescue workers, evacuees who took advantage of these "portable benefits" outnumbered family members. (The Red Cross had also estimated that thirty thousand people would use the benefits over a three- to five-year period, but after two years, only 9,204 people had enrolled, and the numbers were dwindling to one hundred a month.)

As the first anniversary of 9/11 approached, the New York State Office of Mental Health got approval to expand its offerings for people who were still experiencing significant difficulties, either because of debilitating grief or PTSD. This marked the first time that specialized, long-term treatment, known as "enhanced services," would be offered through a federally funded program. According to April Naturale, Project Liberty estimated that about 10 percent of their original 2.5 million projection for people directly affected would fall into this new category, or 250,000 people.

Uncertain of how to identify and treat these people, Project Liberty got in touch with Dr. Katherine Shear, one of Holly Prigerson's colleagues at the University of Pittsburgh, who in September 2001 had published a pilot study on treating thirteen patients for what was at the time being called traumatic grief. The treatment consisted of the following: First, they were audiotaped describing the death as if it was happening in the present with their eyes closed while the therapist helped identify "hot spots," or periods of intense emotion. Participants were then assigned daily homework, which included listening to the audiotape as well as doing exercises to target situations that the patient had been avoiding because it reminded them of the death. Over time, participants were scored on the Inventory of Complicated Grief, a measure developed by Prigerson, and they exhibited marked improvement when compared to a control group that received traditional psychotherapy.

April Naturale wanted Katherine Shear to come to New York City and train Project Liberty's counselors. Naturale recalls Shear telling her "There is no empirical model for working with complicated grief," to which Naturale replied, "Well, we either have you or we have nothing." In the spring of 2002, Shear surveyed 149 Project Liberty crisis counseling recipients to determine the proportion who screened positive for complicated grief. (This is what's known as a convenience sample as opposed to a random sample, because it consists of people already seeking services.) Half the recipients knew someone who had been killed in the attacks. Of those, 23 percent screened positive for complicated grief, and 21 percent exhibited threshold symptoms. Looking at the data another way, the majority of people—56 percent—who both knew someone who died and were already receiving counseling, did not show symptoms of complicated grief a year and a half after the attacks.

That summer Shear taught six hundred therapists to diagnose and treat traumatic grief. (A different specialist was brought in for PTSD.) In the introduction to her training manual, Shear wrote, "It is important to recognize that grief is not a voyage from which we return. Instead it is a permanent state . . . grief is a new homeland." Her treatment for 9/11 victims' family members included having patients keep a grief monitoring diary and an emotions worksheet, and exercises in which they would repeatedly tell the story of the death and engage in imaginary conversations with the deceased. In the end, according to April Naturale, only three hundred people signed up for enhanced services for traumatic grief or PTSD, versus the 250,000 they had anticipated. Even if the vast majority who needed help went elsewhere for it—and there's no way to know how large that number might be—this phase of Project Liberty, while well-meaning in its attempt to prepare for the worst, was enormously wasteful, especially when competing agencies such as the Red Cross were also providing free long-term counseling to families, evacuees, and rescue workers. Project Liberty officially wound down in December 2004, three years after it was launched, and April Naturale is now the managing director of Psychology Beyond Borders.

As for traumatic grief, the terminology reverted to complicated grief after 9/11. "Results showed that it really was very different than a traumatic stress reaction," Prigerson says. "It was not fear-based and it wasn't that survivors were avoiding the life-threatening event. If anything they were stuck yearning for the deceased." Which meant that complicated grief more closely resembled normal or regular grief than experts had thought, and was best defined by the severity and persistence of the reaction rather than the presence or absence of any specific symptoms. And so the name for the disorder was changed yet again—to Prolonged Grief Disorder (PGD)—and the latest version of its definition states that it must last at least six months and create significant functional impairment. Prigerson has forwarded all her data to the mood disorders work group for *DSM-5,* a committee charged with making recommendations for any changes to the manual. "I think it fits most closely with the mood disorders, and they claim to be taking inclusion very seriously," she told me in 2009. The mood disorders group didn't agree and in early 2010, PGD was kicked over to a sub–work group on trauma for consideration. (This isn't a good sign, as a new disorder's chances of making it into the *DSM* decrease without an obvious home.) In March of that year, Prigerson "defended" PGD to the trauma work group over a conference call. "I've received e-mails from them tweaking the criteria we proposed, so I'd like to interpret that as good news," she reported to me afterward.

Of course, when reputations and livelihoods are at stake, there is considerable politicking—both for and against the creation of an official bereavement disorder. Some grief counselors think that adding the diagnosis to the *DSM* will stigmatize grief in general while lining the pockets of psychiatrists, psychologists, and clinical social workers who will have a new diagnosis at their disposal. "This is all about money, making a diagnosis that's pathological so that they can get reimbursed by insurance or get money from the government," one longtime grief counselor who entered the field after the death of her husband (and asked to remain anonymous) told me. It is the less credentialed counselors who stand to lose most, because clients with Prolonged Grief Disorder will be referred upward for reimbursable treatment and because the existence of a diagnosis for complicated grief makes normal grief more benign by comparison (and its treatment less necessary). The grief culture doesn't like dividing grief into categories of severity—or the fact that for the majority of people grief fixes itself—because doing so underscores that only the worst-off can benefit from the help of a counselor.

Even counselors who are supportive of PGD as a distinct entity say that they worry about the long-term implications of including it into the *DSM*. "My biggest concern is how the diagnosis is going to be actually used in the real world, that five years from now we'll be seeing TV commercials with smiley faces and wind-up dolls saying, 'Are you sad because your husband died? See your doctor and we can get you up and running in no time,' " says John R. Jordan, the psychologist who specializes in loss from suicide. "I'm concerned drug companies will see this as a mother lode to be treated by pharmaceuticals instead of psychosocial care."

Prigerson's response to the various critiques is that, as with all disorders in the *DSM,* the goal is not to stigmatize people but to identify and alleviate a condition for which there may even be a biological basis. In 2008, Mary Frances O'Connor, a psychologist at UCLA who studies emotional regulation, recruited bereaved women who had lost a mother or a sister to breast cancer in the last five years. After an initial interview, in which O'Connor and her team administered Prigerson's Inventory of Complicated Grief, half the women were assessed with complicated grief and the other half noncomplicated grief. Participants shared photos of loved ones and discussed their deaths. Researchers then scanned the participants' brains (using a functional MRI) as they were shown the photos and flashed key words from the death stories. It turned out that while the area of the brain associated with pain lit up for both groups, the complicated grief group also had activity in the brain's reward networks. The study's authors suggested that the continued craving of past relationships and the addictive aspect of this neural response prevented people with complicated grief from adapting to their loss. (This doesn't mean that they derive pleasure or comfort from their prolonged pining—on the contrary, it probably makes them feel pretty miserable—just that it may be part of a self-perpetuating feedback loop.)

Not that complicated grief is all just a matter of neurons firing, or misfiring, as the case may be. The list of factors that put people at risk for complicated grief is long and continues to grow. One predictor is the nature of the survivor's relationship to the deceased—was he or she overly dependent on the loved one, or did they see the deceased as an extension of themselves? Certain traits and attitudes may also predispose people to complicated grief: Are they prone to ruminate and do they feel that the world is an unjust place or that bad things always happen to them? A lack of protective resources (money or a steady job) can impede recovery, as can other stressful events. Is the person isolated and lacking the social connections that can provide emotional and practical support, for example? Then there is the survivor's psychic history: Has he or she experienced major losses in the past, or suffered from anxiety or depression before the death? Any previous mood or adjustment problems increase the probability of struggling more intensely after the death of a loved one. And with mood disorders so prevalent in the general population, a survivor's difficulties may predate the loss and do not signal complicated grief at all but, rather, chronic depression.

On the other hand, it's also very possible that the death of a loved one has in fact triggered a major depressive episode, a theory espoused by psychiatrist Sidney Zisook of the University of California at San Diego. According to Zisook, millions of people go undiagnosed and needlessly suffer because of the misconception that depression is a natural feature of bereavement and that medication will blunt emotions and impede necessary grief work. "The natural inclination is to 'normalize' depressive symptoms and consider that under the circumstances, 'anyone would feel down,'" Zisook wrote in *Psychiatric Annals* in 2008. Zisook draws a clear distinction between the experience of grief, where positive feelings coexist with sadness, and the unwavering and pervasive negativity and hopelessness (with possible thoughts of suicide) that clinical depression imposes on its sufferers. He believes that bereavement-triggered major depression is as chronic and debilitating as depression precipitated by other life events (or with no trigger at all), and should be treated early and aggressively with drugs and psychotherapy. Although grief counselors almost invariably discourage medication, aside from perhaps some Ambien to help someone sleep or the occasional Xanax, Zisook and others have run small trials using Wellbutrin, Zoloft, and Lexapro on depressed widows and widowers and found significant improvement in depressive symptoms, along with mild improvement in the intensity of grief.

THE FLIP SIDE: RESILIENCE

It was the attempt to chart a person's well-being prior to becoming widowed that led George Bonanno to his startling discovery of complicated grief's opposite: the much more common, though less examined, phenomenon of resilience. Resilience in bereavement is reaching an acceptable adjustment to someone's death within a relatively short period of time. As Bonanno defined it in 2004 after discovering that about 45 percent of his sample of bereaved spouses showed no symptoms of grief (depression, yearning, despair, anxiety) six months after their spouses died, resilience was the ability "to maintain relatively stable, healthy levels of psychological and physical functioning" despite extremely disruptive events. Resilience wasn't a personality trait, although personality does play a small role. It was a widespread reaction that had gone completely unnoticed—or mislabeled absent grief, which was considered pathological. "Resilience to the unsettling effects of interpersonal loss is not rare but relatively common, does not appear to indicate pathology but rather healthy adjustment, and does not lead to delayed grief reactions," Bonanno argued in an article in the *American Psychologist* in 2004. His groundbreaking studies—the first to gather pre-loss data and also follow bereaved people for four years afterward—have begun to entirely redefine our understanding of bereavement: what has largely been regarded as a catastrophe must now receive consideration as an event that might leave few damaging marks for many, if not most, who experience it.

I first spoke with Bonanno in 2006, after a newspaper article about one of his studies of bereaved spouses caught my eye. (How could it not? The headline read: "Good Grief! Spouses Cope Well.") Bonanno fell into bereavement research by accident when he landed a job running a study in San Francisco just after completing his doctoral work in psychology at Yale. "I didn't know much about bereavement, and frankly I was not enthusiastic. I thought it was a kind of creepy topic to study," he recalled. What did excite him, however, was the realization that there was a dearth of good scientific research on grief. "When I began to read the literature, I was really surprised, stunned even, as to how backward it seemed. The dominant theories about bereavement seemed completely out of date." Then Bonanno came across an article in the *Journal of Consulting and Clinical Psychology* that questioned the common assumptions that severe distress and depression are inevitable after loss, and that grief must be "worked through." The precedent gave him a scholarly green light and the motivation to begin gathering data, in as reliable a way as possible, to support or disprove those assumptions. "Over the years, I have become even more fascinated with the subject because my work has shown that people cope so well—that this is an uplifting story, and not creepy at all," he said.

Bonanno is tall and handsome and has an open and engaging manner. His office at Teachers College of Columbia University is filled, along with a professor's usual books and stacks of papers, with souvenirs from his trips to Asia with his wife, Paulette, who spent a year in Beijing as an exchange student. There's a bit of a Zen quality about him, perhaps acquired from having to defend an apostatic position. When he began publishing his research, the grief culture ignored or attacked it; several years later, it was starting to be acknowledged, albeit reluctantly. In the grief counseling course that I attended during the 2009 conference of the Association for Death Education and Counseling, our instructor did cite

Bonanno's results, but added, "These particular people are very big on resilience, they say we have to stop thinking of grief as being so terrible and were thrilled to have data to support it."

As Bonanno well knows, people have a hard time looking at grief with scientific detachment, and instead imbue it with moral significance. "If you're resilient after a horrible accident or a traumatic event, then you're a hero, but if you're resilient after a death, then you're cold," he points out. This might explain why, even while complicated grief has been around for over a hundred years, adult resilience to loss was only "discovered" in the last decade. (Resilience in children was identified earlier, in the 1970s, when studies found that, contrary to expectations, kids often overcame disadvantages and hardships such as poverty, abuse, or abandonment. The phenomenon was found to be so common that Ann Masten of the Institute of Child Development at the University of Minnesota famously called it "ordinary magic.")

But Bonanno's initial discovery has withstood the test of time, and he's shored up his data by researching what factors can predict a person's resilience to loss. This work has led him to conclude that, contrary to the emphasis that the grief culture has placed on expressing anger and tears, laughing and smiling are more helpful in adjusting to loss, along with the repression of negative emotions. In 2006, he received a $1.5 million grant from the National Institute of Mental Health to run the first-ever experimental studies on bereaved people. Bonanno divided his sample of widows and widowers into two groups: people who had complicated grief and people who were asymptomatic. (He was measuring them at eighteen months post-loss so he couldn't call them resilient. He also had a control group of people who were still married.) Bonanno then ran a series of tests on both groups to try to figure out what distinguished one from the other. For example, he flashed a bunch of differently colored names of colors (the word "blue" might be in blue, or green, or red, or yellow lettering) and asked his subjects to identify the color (not read the word), a difficult exercise known as a Stroop Task, which measures a person's ability to manage his or her attention. Then he would provide them with a subliminal reminder of their loss by flashing the name of the deceased for only twenty milliseconds, long enough for the brain to register it but not so long that the subjects became aware that they were being "primed." Afterward, the people with complicated grief took longer to perform the Stroop Task, which demonstrates, Bonanno says, that the deceased still literally interfered with their cognition.

In another test, Bonanno showed the two groups emotion-evoking photographs. (The images—of injured people, crimes, dangerous weapons, and airplane crashes—had been tested on a large sample in advance and found to reliably induce fear and horror.) He then asked the participants to either show or suppress their emotional reactions, an assignment that allowed him to measure a person's "expressive flexibility" or their ability to modulate emotions. The people with complicated grief had the hardest time turning their emotions off and on, to their detriment. "The more you can modulate emotions, the more you can cope with life events," Bonanno explains.

Although Bonanno is certainly the pioneer of adult resilience, other psychologists are also pursuing the phenomenon. In 2007, Toni Bisconti of the University of Akron explored a particular aspect of personality that, she has written, appears to help women bounce back from widowhood. She surveyed fifty-five widows one month after the death of their spouse and found that those who felt the lowest amount of stress possessed a personality trait called "dispositional resilience," which was defined by three components: they remained connected to other people, rather than isolated; they felt that their grief was manageable and under control; and they embraced and learned from new experiences, rather than avoiding or feeling threatened by them. They were psychologically hardy, optimistic, and able to rise to the challenge that widowhood presented. One widow in the sample remarked, "Well, I have never let anything really get me too far down. My mother had a saying, 'laugh and the world will laugh with you, weep and you will weep alone.' And so I think I've kind of grown up on that theory. I don't think people really want to hear about all your sadness. I've learned that it's probably good to share some of it, because if you don't share, people aren't so likely to share with you." As Bisconti and her colleagues pointed out in an article in 2007, dispositional resilience not only helps people rebuild but also creates an upward spiral effect by attracting friends and family members to their side, who in turn provide more social support. Bisconti went so far as to suggest that grief counselors should not only focus on "alleviating stress (or perceived stress) in older adults, but on promoting resilience as well."

This may prove a tall order. Just as with complicated grief, resilience seems to have many pathways. "Personality probably predicts only about 10 percent of resilience," Bonanno cautions. "Having money helps, having social support helps, having minimal other sources of stress helps, but nothing is a big predictor, it's all just a bunch of things that explain little pieces of the pie. We'd like to have four slices, but it's probably more like twenty slices." It would be pretty difficult for a therapist to induce in a patient a resilient reaction, but that hasn't stopped some from trying. A couple of years ago, Bonanno got a call from the office of the Joint Chiefs of Staff asking him to devise a resilience training program to prepare servicemen for combat and hopefully prevent the high rate of mental problems in soldiers who had been deployed in Iraq. "I told them that I can't devise a program because we don't know enough about resilience yet." So the military called Martin Seligman, the guru of positive psychology and author of *Authentic Happiness: Using the New Positive Psychology to Realize Your Potential for Lasting Fulfillment,* and he devised a program for them instead, which is being required for all

1.1 million of the Army's soldiers at a cost of $117 million. "In medicine, people would never go ahead with a treatment when they don't know enough about the disease," Bonanno told me. "With psychology, the cart is always down the road before the horse. The same thing happened with grief, and now it's happening with resilience."

7

Grief and the Sexes

The death and dying movement was still in its infancy in the early 1980s when its proponents turned their attention to gender and bereavement. Did women and men have different "grieving styles," and whose worked best? Although the answer to the first question was still unclear, the grief culture's overwhelming consensus on the second was that the female response was healthier. The women's liberation movement, for all its accomplishments, was under way, which "enabled not only women, but also their emotionality, to move into the public sphere," as sociologist Tony Walter of the University of Bath has written. Thanatologists and feminists shared the goal of correcting the stoicism and silence that had surrounded death since the 1920s. "Our society's denial of death, guilt, dependence, and vulnerability . . . do not generally describe the experiences or values of women, who usually remain too connected to the lives and deaths of others to participate in this cultural denial of experience," Monica McGoldrick, a family therapist and former president of the Association for Death Education and Counseling, wrote in her book *Living Beyond Loss: Death in the Family.* "The general failure of the literature on death to discuss gender is remarkable, given that the ways men and women handle death are so profoundly different."

But a lack of good data made the discussion particularly vulnerable to stereotype. This was compounded by the emergence of sex essentialism as the dominant school of thought in the behavioral sciences, thanks to Carol Gilligan's bestselling 1982 book, *In a Different Voice,* in which she argued, using slim evidence, that women were hardwired to be more caring and relational. Deborah Tannen's *You Just Don't Understand: Women and Men in Conversation* widened the divide between the sexes by suggesting that not only are our voices different, but we don't even speak the same language. (The gulf was further enlarged by *Men Are from Mars, Women Are from Venus* by John Gray, a former follower of the Maharishi Yogi with a correspondence school doctorate.)

As the grief movement placed more and more emphasis on expressing one's negative emotions, it privileged a response that was stereotypically female. "It's a bias in Western counseling in general," says Kenneth Doka, a professor of gerontology at the College of New Rochelle and a co-author of *Grieving Beyond Gender*. "It begins with the very first thing a counselor or therapist usually says in response to a client, which is 'How does that make you feel?' " The emotive approach presented a problem for a lot of bereaved men, however, particularly widowers of a certain generation. How could they possibly be expected to go through five stages and do their grief work if they had been taught that showing sadness and fear was a sign of weakness? And did this failure to grieve in the prescribed manner mean that they were destined to have psychological problems?

In the end, a place was found for less demonstrative forms of grief. In the 1990s, after he noticed that his own reaction to the death of his father did not fit the prevailing model, Doka, a former Lutheran minister turned thanatologist, divided grievers into two types: those with an "instrumental" style, who responded in intellectual or action-oriented ways, such as a father who shed few tears when his infant son died but spent weeks in his workroom hand-chiseling a stone memorial, and those with an "intuitive" style, who experience grief with much more outward emotional expression. (Doka's schema was and remains purely theoretical, as he has done no surveys to confirm its existence, unlike, say, Bonanno's delineations of resilient, recovered, and chronic grievers, which grew out of methodologically compiled data.)

Despite Doka's attempt to represent the masculine experience of grief (he and a clinical social worker named Tom Golden, who wrote *Swallowed by a Snake: The Gift of the Masculine Side of Healing,* are considered the two experts on male bereavement), men still have a reputation for resisting the overtures of the grief culture. At the Association for Death Education and Counseling's annual convention in 2009, two female social workers presented some ideas for luring men to join grief support groups, such as using activities like golf or playing cards instead of sitting around in a circle talking. (They also advised creating men-only groups, which have the added benefit of preventing intra-group dating. "It happens, and sometimes it ends well and sometimes it doesn't, and if it doesn't end well, then there's the issue of who has to leave the group," one woman who had run coed groups told the audience.)

There's no actual data on whether men join support groups at a lower rate than women, but we do know that the modern grief experience has been observed through a predominantly female perspective. From the very first studies to identify the sex of the participants in the 1950s and 1960s, researchers have focused more on women in general (and widows in particular) than men. "Grief was defined largely as how middle-class English and American widows experienced it," Doka says. The imbalance continues to this day—there are usually far more women in study samples, a phenomenon that's largely logistical: women tend to outlive their mates, so are more plentiful and available as study subjects (and potential support group participants). They also tend to marry men who are older than they are (and who therefore die earlier), and they remarry less frequently than men after a spouse dies, although this may be partially a function of a smaller dating pool. According to the U.S. Census Bureau, as of 2009 there were 13.3 million widows compared to 2.8 million widowers, or a ratio of about 4.7 women for every man.

In 1989, the Dutch psychologists Margaret and Wolfgang Stroebe wondered how the gender imbalance of study samples might be affecting our generalizations about how people cope with grief. They also wondered if the psychic profile of the type of person who agreed to participate in grief studies skewed one way or another. If people who were functioning the best were more willing to participate, the impact of bereavement on the general population could be underestimated; conversely, if people who were the worst off were more willing to participate (seizing, perhaps, on an opportunity to talk about a loss), then bereavement's toll could be exaggerated. So the Stroebes surveyed men and women who had lost spouses but had *not* agreed to participate in a study (but were willing to fill out a questionnaire) along with those who had participated. They found that men who were less depressed were more likely to join a study, but the opposite was true for women. Widows who were *more* depressed were more likely to join, a selection bias that the Stroebes attributed to gender norms for revealing one's emotions. Their results suggested that selection bias alone might be the very root of the stereotype that women grieve harder and longer than men.

But the Stroebes went further. In 2001, they decided to closely examine all studies that had attempted to measure who suffers more, men or women. In order to figure out the answer, two conditions about the research had to be met: First, the widows and widowers in the sample had to be compared to a control group of married women and men. As the Stroebes pointed out, mental distress is not unique to grief alone, and women suffer higher rates of depression in general, regardless of whether they've lost a husband. Second, if the studies didn't have a control group, they had to have evaluated the participants before the loss to establish a baseline of their mental health to be able to state with certainty that their bereavement was the cause of their distress. Any studies that didn't meet either of those conditions were tossed out, and the Stroebes' resulting analysis came to a surprising conclusion: relatively speaking, men suffer more from being bereaved. Yes, bereaved women measured higher on depression scores than bereaved men, but not once women's pre-bereavement or control group depression levels were factored in. Clinicians, and the general public, might certainly be under the impression that there are more widows who are depressed out there, but that's because there are many more widows in general—and therefore more depressed widows to observe.

We also need to look at how much grief is shaped by the

gender beliefs of the people who write about and attempt to help others cope with a loss. In the mid-1990s, two psychologists named Eugene McDowell and Judy Stillion were interested in what death and dying professionals thought about sex differences in grief. So they sent questionnaires to grief counselors and death educators who had been certified by the Association for Death Education and Counseling, and asked them to draw on observations from clinical practice. Fifty counselors responded—thirty-five women and fifteen men, and it was the women who were more likely to declare that their clients grieved in gender-stereotypical ways. While only 27 percent of male counselors said that men and women required different amounts of time to work through their grief (with men needing less time and becoming more quickly involved in hobbies, work, or remarriage), 43 percent of female counselors said that women grieved for longer because they had deeper attachments. When the counselors were asked if they believed that men and women were more alike or substantially different in how they responded to the death of a loved one, 67 percent of the male counselors said "more alike," compared to 49 percent of the female counselors. Only 20 percent of the male counselors said bereaved men and women were "substantially different," while 34 percent of the female counselors said the differences were substantial. In other words, the women were more likely to endorse sex stereotypes about grief than the men. Women are also more likely to become grief counselors in the first place.

*   *   *

John B., whose wife Felicia died suddenly in April 2008, is skeptical of the prevailing stereotypes about men and grief, even while his story might confirm some of them. John and I live in the same town and our children were in the same class in nursery school. About six months after Felicia had died, I asked John if I could interview him, suspecting that his experience could provide me with some anecdotes to illustrate the point that, just as we expect women to grieve indefinitely, we don't allow men to grieve enough. But our conversations over the course of a year and a half served mainly to illustrate my foolishness in reaching for such generalizations.

John and Felicia had met in law school and had both worked in private practice in New York City, but after the birth of their first daughter, they cleaved to an old-fashioned division of labor with Felicia staying home and John becoming the sole breadwinner and working long lawyer's hours. "It got to the point where as I left for work in the morning, my older daughter would say, 'See you tomorrow morning, Daddy!' " John recalled.

One day, when their daughters were three and six, Felicia called John at his office to tell him that she wasn't feeling well and needed help putting the girls to bed. John came home early, and stayed there for the next two days to look after Felicia, who was running a fever and had other symptoms that seemed to add up to a seasonal flu. When the weekend arrived, John noticed that she seemed dehydrated. A doctor friend came over to have a look at her and said, "I don't know what's wrong, but you should get her to a hospital." John left the girls with a baby-sitter to take Felicia to the emergency room. Less than twenty-four hours later, Felicia died of what was eventually determined to be streptococcal toxic shock syndrome, in which a common bacterial infection becomes extremely invasive and causes damage to the vital organs. John had been sent into a waiting room while doctors in the critical care unit prepared Felicia, who had been put into a medically induced coma, for a minor operation at the site of the infection. When he returned, "all of a sudden there was this large group of people around me, and they kind of grabbed me and took me into this room that said 'Family Counseling' on the door, and I thought, *This is* very *bad*."

After Felicia's death, John restructured his life so he could care for his young daughters, R. and L. (to protect the girls' privacy, John requested that I use their first initials). He took six months off from work, and did his best to stick to the routines that Felicia had established in order to provide some semblance of continuity for the girls. When it came to his new role, John feels that he was given a generously large margin of error by the community. "There was sort of the 'soft bigotry of low expectations,' so if I could get them to school looking like they had been bathed in the last seventy-two hours, then I was doing great." He also received help from neighbors and friends, who organized a meal drop at his house for the next six months. Two mothers even deputized themselves as social coordinators for the girls. "They sort of put out the word to other parents that John was a little overwhelmed, and if anyone wanted to schedule a play date, to contact them," he said. He took the girls to a bereavement support group for children where the parents regularly peeled off to talk in private. He found this helpful, and he thought the girls did too.

By the time the fall rolled around, he'd hired a baby-sitter to pick up the children after school and shop for groceries and prepare meals, and was planning to return to work three days a week. Although John had taken out a large life insurance policy for himself, which would have allowed Felicia to stay home until the children were older and not have to return to work if he'd been the one to die, "we did not provide for the other possibility," he said.

When John and I first spoke, in November 2008, he talked about how he had tackled the six months following Felicia's death with the help of a long to-do list. He'd had to delegate his projects at work to colleagues, plan Felicia's memorial service, handle the matters of her estate, pick out her gravestone and plan for its unveiling, and figure out donations he wanted to make in her name and other ways to memorialize her. John and his daughters made it through the winter holidays, their first without Felicia. "I was really focused on making things okay for the girls," he said. "And once L. resigned herself to the fact that you can't put 'a new Mommy' on your Christmas list, because that's not Santa Claus's department, she began to ask her older sister, R., to write some toys down for her."

Still, focusing on busywork—what Kenneth Doka would call an "instrumental" response—hardly prevented him from feeling deeply the loss of his closest companion of the last sixteen years. "It's great having the kids, because they really suck up a lot of time and energy in ways you don't end up minding even though you think you will, and I end up feeling on top of things and don't really have the luxury for other emotions. But then there are these waves when I really, *really* miss Felicia," he said. "I'm functional, but sometimes I have to tell the kids to go watch more TV than is good for them so I can curl up in a corner and cry." Certain situations tended to flood him emotionally. "To hear L. talking about Felicia can make me burst into tears. For example, she'll pick a jelly bean off the floor and say, 'Look, a blue jelly bean! It reminds me of Mommy because blue was her favorite color.'" Another time, while riding the subway to work, he began to cry when he got an e-mail on his BlackBerry with a story about Felicia from one of her friends. "I've discovered that it's possible to burst into tears on the 4 train and have no one look at you strangely," he said.

Overall, though, John felt fairly positive about his ability to handle things on his own. "The analogy I've thought of, and I've been trying to avoid analogies, is that I was in a horrible accident and I'm in a wheelchair, and you can either focus on the wheelchair and how you can get around in it much better than you expected to, or you can say, *God, I really want my legs back*," he said, before adding, "I was never really *not* functional, because I couldn't be. Maybe it's delayed, and will still catch up with me."

But mostly, it didn't, and life in his household grew more stable. The next time we spoke, it was almost two years since Felicia had died. "We've reached an equilibrium," he said. "I'm sort of resigned and fatalistic about it. I'm functional enough from day to day, and don't feel any resentment for 'being abandoned,' but that doesn't mean that I'm happy with the situation." He was still taking one of the girls to the support group (the other had decided to stop going). Felicia's shoes still lined the closet they'd shared, but John had taken off his wedding band several months earlier, and was beginning to think about the possibility of starting a new relationship, and eventually, remarriage.

During his second year without Felicia, John had gotten creative about ways to honor her memory. He decided to host a party around the time of what would have been her fortieth birthday, and played songs that reminded him of her. For the kids, he found himself borrowing rituals from other cultures that he could organize around holidays. "We're trying to come up with different occasions to go to her grave. Since Felicia's family was partially Cuban, we did a sort of post–Halloween/Day of the Dead thing, where I told the girls, 'Take one piece of Halloween candy that Mommy would like.' It turns out that the Russian Orthodox Church (which we have joined although I was not raised in it) has a tradition of visiting the grave nine days after Easter, so we'll take her some eggs then."

John was glad to be able to attend the girls' school events and put them to bed every night, to be the person to comfort them when they needed it. When Felicia was alive and L. woke up in the middle of the night, she'd call out "Mommy, Mommy!" and now she cried, "Daddy!" Occasionally, he'd get comments from strangers who weren't used to seeing such a hands-on father, such as the time he flew back with the girls from Puerto Rico where they had been on vacation with Felicia's parents and a woman told him that he was "brave" for flying solo with two small children. "I didn't say anything about my lack of alternatives," he recalled. "I prefer that kind of comment to the risk of being subtly judged as not sufficiently together."

I asked John if he thought there was a gender divide to widowhood, and he said, well, not much of one as far as he could tell. "I would be very surprised if there were no differences," he said. "Certainly if some segments of society still maintain a masculine ideal about concealing emotion, stiff upper lip, remaining in control, blah blah blah, it's going to affect some percentage of men to some extent, but it's hard to predict that extent and I don't know how easy it would be to quantify."

The problem, as he pointed out—citing the work of Mark Liberman, a linguistics professor at the University of

Pennsylvania—is that statistical differences between groups are much more graspable when shown graphically, usually in overlapping distribution curves. I gave some thought to this and came up with a fictional example of what he was describing:

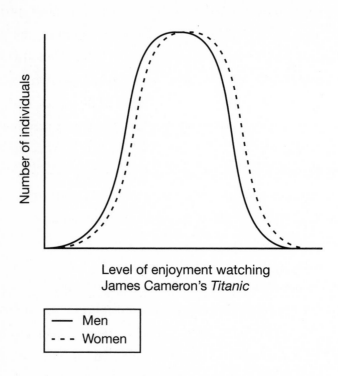

Level of enjoyment watching
James Cameron's *Titanic*

—— Men
- - - Women

As you can see, there is a lot of overlap in how much men and women liked *Titanic*. But the way these kinds of results are described (especially in newspapers, magazines, and morning talk shows) is usually as: "Women enjoyed *Titanic* more than men," which implies that every woman liked the movie more than every man, which is far from the case. As John pointed out, the problem seems to be a lack of language to describe what is essentially a mathematical phenomenon. "It's very hard to talk about statistical differences without lapsing into overly binary or 'essentialist' language," John said. "But that doesn't mean that there aren't some differences."

How great are those differences? It's hard to know because, aside from the Stroebes' review of who suffers more, nobody has done a meta-analysis to determine the average effect size of studies on how long women and men take to grieve. (The effect size is the measurement of the magnitude of an effect, in this case, the magnitude of the sex difference. A meta-analysis takes all the available research findings across many studies of the same question and calculates the mean effect size.) But if meta-analyses of sex differences in other psychological, emotional, and cognitive functions are an indication, much of what we assume to be a significant difference between the sexes is largely inflated. In 2005, the psychologist Janet Shibley Hyde of the University of Wisconsin–Madison reviewed forty-six meta-analyses (sort of a meta-analysis of the meta-analyses) of studies looking at the difference between men and women in a range of traits that included mathematical and verbal aptitude, executive functioning, the tendency toward self-disclosure, helping behavior, self-esteem, and moral reasoning. She found that 30 percent of the effect sizes were close to zero, and 48 percent were small. (Fifteen percent were moderate, 6 percent were large, and only 2 percent were considered very large. The characteristics for which differences between the sexes were moderate, large, or very large were throwing velocity, attitudes about casual sex and incidences of masturbation, and physical aggression.) In her conclusion, Hyde proposed a complete overhaul of our way of thinking about sex differences that she called the Gender Similarities Hypothesis: "Males and females are similar on most, but not all, psychological variables. That is, men and women, as well as boys and girls, are more alike than they are different."

In 2000, Dale Lund, the editor of the book *Men Coping with Grief,* faulted his fellow thanatologists for emphasizing dissimilarities in the grief response without looking at hard science. "For many years, we have relied on cultural beliefs, practices, norms, values and oftentimes stereotypes to answer questions about gender differences," he wrote. "We are becoming so narrow in our 'search for differences' that we are ignoring much of what we all have in common." Lund had conducted several longitudinal surveys of widows and widowers over the preceding two decades, and concluded that men and women shared the same range of emotions at similar levels of intensity and with similar frequency. Even the changes over time were alike. "This finding does not mean that men experience the death of their spouse in *exactly* the same manner as women do, but there is evidence that their adaptation follows much the same course over the first two years according to the global adjustment measures used in this analysis."

Another problem with searching for sex differences is that they are susceptible to a phenomenon called stereotype threat, which has been amply illustrated by the pioneering work of social psychologist Claude Steele at Stanford. In a classic example of stereotype threat, male and female college students with equivalent math backgrounds were recruited to take a math test. When participants were told that the test they were about to take had shown sex differences in the past, women scored lower compared to men. When participants were told that men and women had performed equally on the test that they were about to take, there were no differences in their scores. Unfortunately, anticipating bias turns out to be a self-fulfilling prophecy. Another phenomenon known as deindividuation, which is an eggheaded way of saying whether someone feels anonymous or not, can even reverse sex differences. For example, in one study, men showed more aggressive tactics while playing a video game (they dropped more bombs on their imaginary opponents) when they thought they were being watched. When participants thought they *weren't* being watched, there were no significant sex differences, and in fact, women dropped a few more bombs. Anonymity appears to neutralize certain sex differences in the grief response as well. In 2007, three researchers analyzed bulletin board postings on the Web site for Compassionate Friends, an international organization devoted to helping parents who have lost a child, and found that bereaved fathers, when provided with the veil of anonymity, were no less emotive or expressive of their love than bereaved mothers, or wrote posts that were more intellectual or indicative of an "instrumental" style of grief.

Back to the question, then: Do men and women grieve differently? The research is too flawed to tell, and if there are differences, they're certainly not as great as the similarities. For the most part, the attempt to "gender" grief—just as marshalling it into stages—does it a disservice. As John B.'s experience illustrates, trying to line up items on a checklist of sex characteristics is a futile exercise: for every detail that seems "male"—he took off his wedding ring after a year and a half, indicating that he moved on quickly—there is a detail to contradict that conclusion, such as the fact that he hadn't removed Felicia's clothes from their closet. Whether someone is a man or a woman has little predictive power about how he or she will adjust to bereavement. To view grief through a framework of gender is more likely to obscure than clarify.

8

Grief for Export

"Americans are the largest producers of psychological research. The overwhelming subject of the research is Americans. The United States constitutes less than 5% of the world's population. Therefore, on the basis of a sample of less than 5% of the world, theories and principles are developed that are mistakenly assumed to apply to human beings in general; that is, they are assumed to be universal."

—Stanley Sue, professor of psychology, psychiatry, and Asian American studies, University of California, Davis

Although our model for grief grew from American soil and adapted to fit our own historical moments, it has traveled far beyond our borders to become part of the worldwide spread of our definition of mental health. As Ethan Watters pointed out in *Crazy Like Us: The Globalization of the American Psyche*, "We are engaged in the grand project of Americanizing the world's understanding of the human mind." The exportation began when Kübler-Ross's *On Death and Dying* was translated into foreign languages, sparking thanatology movements in parts of Europe. Kübler-Ross was invited to conduct workshops abroad, and her former employees continue to teach her "externalization" technique in Ireland, Germany, Spain, and New Zealand. In Australia, the Kübler-Ross model won immediate acceptance. "Since [the publication of *On Death and Dying*], it has been an article of faith among both American and Australian health professionals that talking about loss is important to resolving it and prevents distressing psychiatric morbidity," Allan Kellehear, the Australian sociologist, wrote in 2005. "Australian ways of grieving . . . are not logical outcomes of our local experience but are rather socially constructed ways of understanding inherited from a variety of dominant foreign influences."

As our model for grief developed an international following, so did our predilection for professional help. In Japan, grief counseling is now routinely offered through hospices, thanks largely to Alfons Deeken, a German Jesuit priest who went to Tokyo as a missionary in the 1960s. In 1982, Deeken founded the Japanese Association for Death Education and Counseling, which organizes memorial services for family members that include lectures by association members on the subject of grief. "The speaker focuses on the loss experience, the typical feelings of survivors, the grief work to be done, the potential difficulties awaiting them on the road to recovery, and the help they can give each other by sharing experiences and by supporting each other," Deeken explained in 2005. The association now has more than forty chapters and even offers "pre-widowhood education" to married Japanese women.

In Hong Kong over the last decade, grief counseling was also introduced through hospice care, and there are now two centers dedicated to guiding the bereaved through their grief work and to training health professionals and volunteers for the task. "There is no indigenous Chinese model of grief therapy, so most approaches are adopted from Western models," Samuel M. Y. Ho of the psychology department at the University of Hong Kong explained to me. One of the first grief support groups in Hong Kong, offered in 1995, encouraged Western-style self-disclosure as the key to emotionally adapting to loss. According to the group's facilitators, who were both on the faculty of the University of Hong Kong, "Sharing of feelings is crucial for healing and personal growth."

If American guidelines for grief have narrowed our own repertoire of responses to loss, they will prove even more constricting to the rest of the world. In 1976, Paul C. Rosenblatt, a psychologist at the University of Minnesota, surveyed seventy-eight different ethnic groups, from the Basque to the Balinese, the Masai to Marshallese, and documented an astonishing diversity of who grieves, how they grieve, and for how long. Among the Matsingenka of Peru, for example, it is dead people, not the living, who are thought to grieve most acutely.

Among the Achuar in eastern Ecuador, the bereaved try to forget the dead, who are seen as intensely lonely and trouble-some spirits, by not using their proper names and referring to them only by pronoun. "My opinion is that nothing—no feeling, no meaning or understanding—related to grief is universal across cultures or probably even within a culture," Rosenblatt told me in 2009.

None of the characteristics that we think of as intrinsic to grief can be generalized to the rest of the world—not even crying. "Someone cries in most cultures when someone dies, but not everyone cries, and the crying has diverse forms, reasons for crying, and meanings from culture to culture," Rosenblatt said. "Fear and anger are less common, and where there is fear and anger, it varies a lot in meaning, forms, and reasons from culture to culture—and again there seem often if not always to be differences within a culture."

The range of responses has to do not merely with differ-ences in religion and custom, but perhaps more importantly, a difference of beliefs about psychological health. In 1988, a Norwegian anthropologist named Unni Wikan published a years-long study of bereavement in Egypt and Bali—both predominantly Muslim countries—and found stark contrasts between the two. When a loved one died in Egypt (Wikan observed mostly poor families in Cairo), relatives screamed, wailed, and lamented for several weeks, venting their suf-fering and pain in front of visitors who both consoled and encouraged the outpouring. According to Wikan, these

Egyptians believe that unhappiness must find its way out of the body or else it weighs on the soul and causes mental illness, a perspective that has similarities to the American school of thought.

By contrast, in Bali, suffering "must be surmounted to allow a person to survive," Wikan wrote. So when neighbors and friends congregated after a death, their goal was not to help survivors express their sadness, but combat it. One Indonesian man who had spent the night at the house of a newly widowed friend told Wikan, "We talked about what had happened, about the sickness and the circumstances of her death, but we laughed and joked a lot to make the family's hearts happy from sadness. We say things like 'what is dead is dead' and 'let bygones be bygones' to remind them that they must not be sad." Too much sorrow weakens the spirit, brings confusion, and makes a person vulnerable. And because suffering is contagious, it is also in a community's interest to try to cheer up the bereaved: "If we express our sadness, we will make others sad as well. It will be bad for all, and dangerous too," the man explained. As a result, Wikan concluded that, even though the Balinese do cry a little bit upon death, they also see laughing as a sensible, protective measure to keep one's psyche strong.

Within this context, cracking jokes to a friend whose wife has just died doesn't seem quite so strange. And yet Americans are often puzzled when they encounter approaches to grief that feel radically different from their own. In 2008,

New York Times columnist David Brooks traveled to Sichuan province in China, where an earthquake had killed seventy thousand people three months earlier. Brooks stopped in the village of Pengshua and interviewed a man who had escaped his collapsing house one step ahead of his wife, who did not. He interviewed another man who was working in his garden when his six-story apartment building caved in and killed his eighteen-year-old son. Brooks was perplexed when both survivors told their stories in what he described as a "matter-of-fact" or "pragmatic" tone. "We'd visited the village without warning and selected our interview subjects at random, but some of the answers were probably crafted to please the government," Brooks wrote. "Still, there was no disguising the emotional resilience and intense mutual support in that village. And there was no avoiding the baffling sense of equanimity. Where was the trauma and the grief?" Brooks went on to characterize his interviews with earthquake survivors as "weird" and "unnerving," before wondering if their ability to remain so optimistic was "emotionally sustainable or even healthy."

After I read Brooks's critical appraisal, I consulted the literature on Chinese bereavement and found several compelling explanations for the equanimity he found so baffling. Research psychologists at the University of Hong Kong, for example, have written that a variety of philosophies popular in China (Confucianism, Taoism, and Buddhism, mainly) emphasize respecting the deceased and easing their transition to an afterlife rather than focusing on the personal impact of the loss. One study of a small group of Chinese widows showed that suppressing one's own emotions had an important purpose: too much sorrow impeded the deceased's spirit, so it was not uncommon to find Chinese families who appeared calm when they gathered around a loved one's deathbed. One widow said that she didn't cry because she feared it would affect her husband's path to his next life. "I even asked him to go without worrying about us. I tried my best to make him leave at ease," she said.

This made me wonder what happened when Chinese immigrants encountered the folkways of American grief. As of 2008, there were almost two million people living in the United States who had been born in China and Taiwan, and several million more American-born people of Chinese ancestry. As a larger group, Asian Americans have a reputation for not cottoning to Western notions of psychological health—one survey that compared attitudes of Asian American and Caucasian American college students found that Asian Americans were more likely to believe that psychotherapeutic probing *caused* mental illness. "It is not that Asian-Americans were unwilling to discuss their problems and secrets openly," the researchers who conducted the survey concluded. "Rather it is the *dwelling* and *in-depth* analysis of morbid thoughts which may be perceived to be detrimental." Western psychology tends to stigmatize such differences. "Asian Americans are viewed as people who are unable to express themselves and don't use mental health services, but that doesn't take into account the things they *do* do, whether through faith or their own coping styles or indigenous healing practices," says Christine J. Yeh, a psychologist at the University of San Francisco. As Yeh has pointed out, while Asian Americans are often characterized as "avoidant," they are often, in fact, coping in culturally appropriate and healthy ways.

Yeh had the opportunity to explore those ways after 9/11, when 184 Asian Americans were killed by the attacks on the World Trade Center. Nine months later, she and her colleagues interviewed eleven of their surviving relatives. Although the sample was small, Yeh found that these immigrants had held on to coping strategies from their native countries. (Participants had been born in China, the Philippines, and Korea.) While they sought comfort within their families or their own ethnic group, they also described not wanting to burden others with their problems or cause unnecessary pain. So they tried to keep their feelings to themselves, stayed active, and focused on logistical matters. Almost all of them saw death as a part of the natural order, or God's will, and that determinism seemed to help their adjustment. As one Chinese man said, "Chinese people's fatalism is a good alternative to counseling. It is like you just have to accept it. Can you fight with your fate? What's the use of fighting with it, if this is his fate?"

Even though free counseling was widely available

through Project Liberty and other agencies, only three of the eleven family members saw a counselor, and those who did found it unhelpful and dropped out immediately. The rest said that they wanted to solve their problems on their own, and worried about bringing embarrassment to their family. Yeh described this orientation as "collectivistic," because family and community took priority over the needs of the individual. "The Western tradition of seeking help from a stranger such as a psychologist may be culturally inappropriate from a collectivistic perspective," she wrote. "In addition to seeking help and support through their connection with important others (friends, family, etc.), Asian Americans may also withhold or forbear their problems in order to maintain social harmony within a close system."

For several days in the spring of 2010, I headed into New York City's Chinatown to see what I might discern of a specifically Chinese American response to loss. (Although that community obviously doesn't represent all Chinese immigrants, it struck me as a good place to take a crude ethnographic measure.) On my first visit, I walked along Mulberry Street, where three Chinese funeral parlors line one city block. I had left a message for the director of one of them in advance, but he never returned the call, and it seemed inappropriate to enter uninvited in case a funeral was in progress. Wedged between a Dunkin' Donuts and a tenement building was an intriguing store called Fu Shou Funeral Products, which sells miniature paper replicas of houses and cars meant to be burned at funerals and subsequent ceremonies to ensure an afterlife for the deceased that is filled with their favorite creature comforts from this world. (The custom comes from an old practice in China of burning spirit money, or joss paper, for dead relatives.)

I was on my way to the Charles B. Wang Community Health Center on Canal Street to talk to Teddy Chen, a clinical social worker and the center's director of mental health. Chen runs something called the Bridge Program, which tries to overcome the stigma of psychiatric services among Chinese Americans by integrating them with primary care. The idea is that people from the community who come in for physical complaints are also screened for psychological problems and referred to Chen or another mental health clinician for more evaluation.

In his office, I asked Chen if the Bridge Program offered any bereavement counseling, and he was almost amused by the question. "Death and dying for Chinese Americans is really, *really* private," he said. "I don't see them coming in here saying, 'I want to talk about my loss.'" Chen described one of his clients, a man whose daughter had died on 9/11 but who had only recently begun talking about it. "The loss is so private, he won't really volunteer it, but he's dealing with it according to his own way," he said. Although Chen was born in Taiwan, he had obtained both his master's degree and

Ph.D. in social work at Columbia University in New York City. "I'm pretty Westernized from my training and education, and I personally believe that people can benefit from discussing the death of a loved one, but I question whether that applies to everyone," he said. Trying to persuade his clients to open up about their losses would be intrusive and counterproductive; instead his goal is first to gain their trust so that they will come back, and then simply provide them with support. For this population, he said, "mental health services can be helpful, but not in a Western way."

There were no programs for the bereaved either at Hamilton-Madison House, another community center that serves Chinatown's population. "Grief counseling would be a strange concept for Asian families, because death is seen as a part of life, and enduring hardship is a virtue that's good for you," Peter Yee, the assistant executive director for behavioral health services, told me. "When we do deal with loss, we don't call it grief counseling, and for the most part we don't deal with the minutiae of their feelings but try to make them stronger." To characterize the community he served, Yee invoked an ethos within which duty to family and a desire for harmony subordinates the needs of the individual. "In the West, if someone dies young or commits suicide, people say *How could they do this to me!* In Chinese culture, it would be rare to get angry at someone for dying, to think *He died but it's all about me now,*" Yee said. "You just have to accept the death and pay honor to the deceased. It's a cultural way of defending our psychic fractures. I wouldn't want to put judgment on whether one approach is better than another, but from a clinical point of view, I think it works." Like Chen, Yee was trained in the United States, but had no interest in converting his clients to a Western model of therapy. "You don't want to go the route of changing someone's whole worldview," he said. "It might make *you* feel better because it's in line with the principles that you learned in school, but it won't necessarily make other people cope better."

If Chinese Americans in Chinatown (who, admittedly, may have arrived more recently and are therefore less Westernized than older generations) don't seem to make use of our secular, therapeutic approach to grief, I wondered if they were drawing support from more spiritual sources. At the suggestion of a psychologist who worked with the hospice team of the Visiting Nurse Service, I decided to visit the Grace Gratitude Buddhist Temple, housed in a narrow, five-story building on a busy stretch of East Broadway, flanked by a Bank of China ATM machine on one side and the QQ Bakery on the other. Half a block away, the B train emerged from an underground tunnel every five minutes to make its noisy journey over the Manhattan Bridge to Brooklyn. Founded in 1968, the temple espouses a branch of Buddhism popular in China known as Pure Land, so named for the place where one goes after achieving spiritual enlightenment or Buddhahood and escaping the cycle of Samsara, or death and rebirth.

A monk by the name of Shih Benkong had offered to

show me around. I'd taken an instant liking to him over the phone because when I asked if he would talk to me about the Buddhist approach to grief, he immediately said, "How about tomorrow? I can do tomorrow. Writers are always on deadlines, right?" Benkong, fifty-eight, was tall with a shaved head and light blue eyes. He wore a light gray cotton tunic to his knees and matching pants, and around his neck hung a beaded eyeglass chain attached to silver wire-rim glasses. Before he became a monk, his name had been Harold E. Lemke and he was born and raised in nearby Jersey City and was fascinated by Chinatown as a child. In the 1970s, Lemke dropped out of his local high school to go to Taiwan, where he studied Mandarin as well as Chinese history and philosophy and completed his diploma. He returned home, majored in Asian Studies at Seton Hall University, then went back to Taiwan and was lucratively employed as an interpreter when he realized he was bored and decided to go to South Africa to help with the AIDS crisis. "And if that kind of work doesn't put you on a religious path, I don't know what does," he said. In 2001, having found a local temple and a master who would agree to tutor him, he became a novice monk and was ordained in Taiwan in 2003. He was branded with three dots on the top of his head to signify that his vows were embedded into his body.

"In Buddhism, death is all about the person dying; we have to help them die as best that they can," Benkong explained, after leading me upstairs to a meeting room on the second floor of the temple, where, with the aid of a dry-erase board, he proceeded to map out the different realms of Buddhist existence. (Which realm you are reborn into depends on your actions during your life, or your karma.) When a member of the temple is dying, Benkong pays a visit and chants sutras to help them move from one life to the next, a transition known as the Bardo state, which lasts for forty-nine days. (This is the time when, in Tibet, Buddhist monks recite their famous book of the dead, *The Great Book of Natural Liberation Through Understanding in the Between:* "Hey, noble one! Listen unwavering with intense concentration! Now you have arrived at what is called 'death.' You are going from this world to the beyond. You are not alone; it happens to everyone. You must not indulge in attachment and insistence on this life. Do not lust! Do not cling!")

Then Benkong led me into a pleasant, sunny room full of plants and statues of the Buddha. Arrayed in long rows on the walls were 913 rectangular, yellow laminated placards with Chinese characters on them. Each bore the name of a person who had died, as well as the phrase *May the light of the Buddha bring them to the Pure Land and create rebirth.* This was where family members came to bring food or flowers or light incense for their loved ones, Benkong explained, scooping up a bunch of bananas someone had left to bring down to the kitchen.

When I asked Benkong if there is anything in the scriptures that addresses bereavement, he said, "Not so much how to deal with grief, but what we can learn from death, which is to realize your own impermanence. The idea is to remember, *As you are losing this person, so are you going to lose everything in life: your job, your kids, your body*. That doesn't mean that you shouldn't care about any of those things, but that you should cherish them while you have them."

The Buddhist system turns the death of others into a powerful incentive to do good in one's own life by embracing positive attitudes and actions. "Not from weeping nor from grieving will anyone obtain peace of mind; on the contrary, his pain will be the greater and his body will suffer," the Buddha said. "He will make himself sick and pale, yet the dead are not saved by his lamentation. People pass away, and their fate after death will be according to their deeds. . . . He who seeks peace should draw out the arrow of lamentation, and complaint, and grief. He who has drawn out the arrow and has become composed will obtain peace of mind; he who has overcome all sorrow will become free from sorrow, and be blessed."

After spending several years researching Western bereavement, it was refreshing to see how Chinese Americans were able to take the sting out of grief while still recognizing the significance of death. This is not to suggest that they or any other ethnicity or religious group necessarily do grief "better"—there are dozens of approaches and philosophies, and this is but a brief look at one. But we have something to learn from the cultural relativity of loss. Different beliefs about grief and alternative modes of healing have value because they remind us that no one particular approach is more effective, or healthier, than another. The exportation of the American model, while done with the best intentions, threatens to erase this diversity. If our model were to become the worldwide standard, the same mistake we've already made at home would be repeated on a global scale, which is to turn a description of one way to grieve into a prescription for the only way to grieve.

Afterword

With this book I hope to offer you a means of escape from our habitual ways of thinking about grief. I specifically chose examples of people who showed resilience not only because they challenge the reigning ideology and help prove a point (and also represent the experience of the majority of people), but because I believe that their stories of putting their lives back together can help others. Perhaps just the knowledge that our survival instinct is strong, and that a great many have not only endured terrible losses but have also thrived, can be a source of hope, something that I found to be quite scarce in our grief culture.

As a society, we will most likely be unable to face grief without some kind of script. "No culture before has abandoned all recommendations as to how to mourn," sociologist Tony Walter writes in *On Bereavement*. But it certainly seems time to move beyond our current habit of using untested theories to create unnecessarily lengthy and agonizing models for loss, ones that I believe have created more fear of and anxiety about the experience. Instead of rushing to prescribe ways to grieve, it would be more helpful to update ourselves on what the little science that has been conducted in this area tells us: that most people are resilient enough to get through loss and reach an acceptable level of adjustment on their own. A smaller minority will have a much harder time of it, and clinicians should focus their efforts on tailoring interventions for this group that are based on evidence, not myth. Even if Prolonged Grief Disorder doesn't make it into the *DSM-5,* drawing a clearer distinction between the severity and duration of common and unusual reactions might ultimately encourage those who don't need outside help, and better help those who do.

We also need to give much deeper consideration to how grief is created and shaped by our social and cultural environments. A better appreciation of just how relative grief can be will make us more aware of the variability even within our own culture and give us more freedom of choice instead of having to hew to one approach because it happens to be the dominant one.

In the meantime, most of us will just have to find our own way, using what works and discarding what doesn't. After Sarah White Bournakel's husband was killed by a drunk driver on Maui, she first looked for guidance from the obvious sources, the ones she had been taught to consult. "I had actually taken a class on death and dying in college, but when I went back to look at Kübler-Ross again, it was totally not helpful," she recalled. Her sister, a school psychologist, gave her a grief wheel, which takes you through shock, protest, disorganization, and reorganization, and Sarah dutifully put it on her refrigerator, but as she pointed out, "It's a wheel, it goes around and around, but how do you get off the damn thing?" People kept leaving grief books on her doorstep. "I had a newborn baby, so I had no time to read. A lot of them were crap so I finally dumped them all." She did make it through Joan Didion's account of widowhood, but said, "It seemed like she was in a fog when she wrote it—I thought, *this is a woman who needs some perspective*." She joined a widows support group, but it depressed her to see women who had been coming for three, four, five years. "It made me feel as though, in addition to losing Stefan, I was condemned to a life of sadness and lonely connections with random people who seemed stuck."

Sarah began to dabble with alternatives. "You name it, I did it—reiki, meditation, I even had an intuitive come to my house to clear it of 'bad energy,'" she said. She set up a shrine to Stefan in her bedroom, a box filled with letters on top of which she placed photos and a Buddha figurine and some Christian icons. She left flowers at the site of the car crash every week, and even saved a bone from her husband's cremation and then went diving off Maui with the leatherback turtles to bury it in the sand. Sarah also turned her attention toward prosecuting the drunk driver who had killed Stefan. She got in touch with a victims advocate from MADD (Mothers Against Drunk Driving), who helped her navigate the legal system and accompanied her to hearings. "They were like a lifeboat," she said. But a year after Stefan's death, she was suffering from recurring violent nightmares and insomnia, so she went to see a psychologist who specialized in trauma. "I was a single parent and clearly exhausted, and a combination of his weekly therapy, massage, acupuncture, and exercise led me back to relatively normal behavior," she said. Sarah also met and began dating the man who would become her new husband, and two years after Stefan's death, they got engaged. "Stefan died happy, and because of that, I was able to connect with someone else, and fall in love with someone else, because I had such a strong connection to Stefan," she said. "You have to find what works for you, and after years of trying everything, I found what worked for me. And it was not in a book."

Acknowledgments

This book was inspired by the work of two researchers, George A. Bonanno of Teachers College at Columbia University, and Holly Prigerson, now at the Dana-Farber Cancer Institute. It was one of George's studies of widowhood that sparked my initial interest in this subject in 2006, and Holly's scrutiny of the stages of grief provided the incentive to expand my efforts from a magazine article to a book. Both were very generous with their time and material; their research is the foundation on which this book stands, and George has since published his own excellent book on grief and resilience, *The Other Side of Sadness*. I also owe a great conceptual debt to Tony Walter, a sociologist at the University of Bath, who got me thinking about grief from a cultural perspective and from whom I have liberally borrowed the phrase "the grief culture."

Acknowledgments

I wish to thank the following people for sharing their insights with me: Janice Genevro, Joseph Currier, Robert Neimeyer, Paul Rosenblatt, Kenneth Doka, Vanderlyn Pine, Colin Murray Parkes, Julie Ann Wambach, Toni Bisconti, Suzanne Degges-White, Allan Kellehear, J. William Worden, John R. Jordan, Chris Feudtner, Christine J. Yeh, Pamela Yew Schwartz, Teddy Chen, Peter Yee, Shi Benkong, Richard Schulz, Peter Nevraumont, Frank Lewis, Frank Lawlis, Phil Anderson, Ryoko Dozono, David Kessler, Alan Wolfelt, C. Knight Aldrich, John Rea Thomas, Barbara Higgins Cox, George E. Dickinson, Karen M. Seeley, Michael Cohen, April Naturale, and Sandro Galea. A great many of the aforementioned work directly with the dying and the bereaved, and I benefited enormously from their wisdom and experience.

Others whose writings and ideas were extremely helpful and to whom I owe credit: Margaret and Wolfgang Stroebe, Henk Schut, Roxane Cohen Silver, Camille B. Wortman, Kathleen Lalande, Vivienne Cass, Beverly Raphael, Phyllis Silverman, Helena Znaniecki Lopata, Robert DiGiulio, Therese Rando, Catherine Sanders, Susan Zonnebelt-Smeenge, Robert DeVries, Dale Lund, Bert Hayslip, Stephen Shuchter, Laurel Hilliker, Sally Satel, Katherine Shear, Stanley A. Murrell, Mary Frances O'Connor, Eugene McDowell, Judy Stillion, Janice Shibley Hyde, Unni Wikan, Stanley Sue, Nigel P. Field, Eval Gal-Oz, Samuel M.Y. Ho, Thomas Schlereth, Peter N. Stearns, Geoffrey Gorer, Philippe Ariès, Carol Acton, Marcia Seligman, and Ron Rosenbaum.

I also wish to thank Valerie Frankel, Sarah White Bournakel, Maggie Nelson Burchill, Alisa Torres, and John B., who all spoke to me at length about losing a spouse. Ray Schmuelling and the other widows and widowers of Young Widow.net graciously allowed me to tag along on their Widowbago weekend. Valerie Molaison tolerated my presence in her grief counseling class, and Joyce M. Lennon and Mina Gates tolerated my questions about why they wanted to become counselors. Jon Radulovic at the National Hospice and Palliative Care Organization, Jessica Koth and Emilee High at the National Funeral Directors Association, and Emily Burch and Gordon Thornton of the Association for Death Education and Counseling all provided me with essential information.

My agent David McCormick immediately understood what I hoped to accomplish with the book and contributed his own clever term, "stageism." My editor Priscilla Painton skillfully guided me through the writing of it, cheering me on the entire way. Michael Szczerban and Gypsy da Silva at Simon & Schuster sprang into action to polish the manuscript, enlisting the expertise of Fred Chase and Jim Stoller. Alex Postman, Anya Sacharow, Diana Donovan, and Tara Bray Smith read parts or all of the manuscript, asked important questions, and pushed me in the right direction. Annik LaFarge gave me sage advice on building a website. Laurie Abraham edited the article in *Elle* that started me on this path. Patricia Perito and the entire staff of the Pelham Public

Acknowledgments

Library gave me a home a block from home for three years; the MacDowell Colony gave me paradise for three weeks

There is no way I could ever have started this book, much less completed it, without the help and generosity of my family. Eleanor and Hugh Sackett, Rodman Davis, Harvey and Marilyn Konigsberg, Henry and Belle Burden Davis, and Townsend Davis and Bridget Elias all provided food, shelter, babysitting, emergency checks, and encouragement. Cindy Anderson kept everything running smoothly and talked me down from a ledge or two. Finally, my husband, Eric Konigsberg, applied his talents to the manuscript not once but twice and supported me in innumerable ways before, during, and after the writing of the book. My biggest debt of gratitude is owed to him.

Notes

INTRODUCTION: THE IDEA THAT WON'T DIE

Page

1   *Lanny Davis, a die-hard Clinton supporter:* Kate Zernicke, "Can You Cross Out 'Hillary' and Write 'Sarah'?," *The New York Times,* August 30, 2008.

2   *After all, his team of three hundred collectors:* David Streitfeld, "So You're Dead? Don't Expect That to Stop the Debt Collector," *The New York Times,* March 4, 2009.

2   *Conan joked on his subsequent comedy tour:* Erik Pedersen, "Conan O'Brien's Legally Prohibited from Being Funny on Television Tour," *The Hollywood Reporter,* April 13, 2010.

2   *Frank Rich has used them:* Frank Rich, "The Petraeus-Crocker Show Gets the Hook," *The New York Times,* April 13, 2008.

3   *The painting sold for:* Web site for Sotheby's London, http://www.sothebys.com/app/live/lot/LotDetailPrint able.jsp?lot_id=15947338, accessed 1/11/09.

4   *Counseling for grief, though well-intentioned, does not:* Joseph M. Currier, Robert A. Neimeyer, and Jeffrey S. Berman, "The Effectiveness of Psychotherapeutic Interventions for Bereaved Persons: A Comprehensive Quantitative Review," *Psychological Bulletin* 134, no. 5 (2008): 648–61. In 2007, psychologist Scott Lilienfeld of Emory University placed grief counseling on a list of treatments that cause harm in an article in *Perspectives on Psychological Science,* but a subsequent re-analysis of the data on which this claim was based found no evidence of "deterioration effects." See William T. Hunt and Dale G. Larson, "What Has Become of Grief Counseling: An Evaluation of the Empirical Foundations of the New Pessimism," *Professional Psychology: Research and Practice* 39, no. 4 (2007): 347–55.

5   *rooted in the principles of psychotherapy:* In *The Triumph of the Therapeutic* (New York: Harper & Row, 1966), Philip Rieff argued that the analytic attitude created a negative community of people sinking more and more deeply into the self.

5   *In 1984, an Institute of Medicine report:* Nancy S. Hogan, Daryl B. Greenfield, and Lee A. Schmidt, "Development and Validation of the Hogan Grief Reaction Checklist," *Death Studies* 25, no. 1 (2001): 1–32.

6   *Sociologists (and state highway officials):* Ian Urbina, "As Roadside Memorials Multiply, A Second Look," *The New York Times,* February 6, 2006. See also *Spontaneous Shrines and the Public Memorialization of Death,* edited by Jack Santino (Palgrave Macmillan, 2006).

7   *A 2002 survey of the guest book entries:* Pamela Roberts and Deborah Schall, "Hey Dad, It's Me Again: Visiting in the Cyberspace Cemetery," presented at the Death, Dying and Disposal Conference, Bath, England, September 2005.

7   *"Telling your story often":* Elisabeth Kübler-Ross and David Kessler, *On Grief and Grieving* (New York: Scribner, 2005), p. 63.

8   *As Tony Walter, a British sociologist:* Tony Walter, *On Bereavement* (Philadelphia: Open University Press, 1999), p. 125.

8   *"Any natural, normal human being":* Marcia Seligson, "Playboy Interview: Elisabeth Kübler-Ross," *Playboy,* May 1981.

9   *But the resulting study, published:* Paul K. Maciejewski, Baohui Zhang, Susan D. Block, and Holly G. Prigerson, "An Empirical Examination of the Stage Theory of Grief," *JAMA* 297, no. 7 (2007): 716–23.

10   *"What might explain the sustained":* Holly G. Prigerson and Paul K. Maciejewski, "Grief and Acceptance as Opposite Sides of the Same Coin: Setting a Research Agenda to Study Peaceful Acceptance of Loss," *British Journal of Psychiatry* 193 (2008): 435–43.

10   *Skepticism of the stages:* Richard Schulz and David Aderman, "Clinical Research and the Stages of Dying," *Omega* 5, no. 2 (1974). See also Anne M. Metzger, "A Q-Methodological Study of the Kübler-Ross Stage Theory," *Omega* 10, no. 4 (1979): 291–300.

11   *a 2008 survey of fifty hospices in Canada:* Hospice Association of Ontario Grief and Bereavement Support Survey, available at http://www.hospice.on.ca/conference/presentations/HAO%20Grief%20and%20Bereavement%20Survey%20Results.pdf, accessed 12/08.

11   *"Stage theories of grief have become popular":* Janice L. Genevro, Ph.D., "Report on Bereavement and Grief Research," Center for the Advancement of Health, November 2003.

13   *Bonanno and his colleagues tracked:* George A. Bonanno, Camille B. Wortman, Darrin R. Lehman, Roger G. Tweed, Michelle Haring, John Sonnega, Deborah Carr, and Randolph M. Nesse, "Resilience to Loss and Chronic Grief: A Prospective Study from Preloss to 18-Months Postloss," *Journal of Personality and Social Psychology* 83, no. 5 (2002): 1150–64.

15   *recently bereaved individuals who did* not *express:* George A. Bonanno, Karin G. Coifman, James J. Gross, and Rebecca D. Ray, "Does Repressive Coping Promote Resilience? Affective-Autonomic Response Discrepancy During Bereavement," *Journal of Personality and Social Psychology* 92, no. 4 (2007): 745–58.

15   *probably the most accurate predictors:* Louis A. Gamino, Kenneth W. Sewall, and Larry W. Easterling, "Scott and White Grief Study: An Empirical Test of Predictors of Intensified Mourning," *Death Studies* 22, no. 7 (1998): 333–58.

15   *"If one has always met life's problems with strength and assurance":* Edgar N. Jackson, *You and Your Grief* (New York: Channel Press, 1961), p. 10.

16   *to borrow a term from two pioneers:* Roxane Cohen Silver and Camille B. Wortman, "The Myths of Coping with Loss," *Journal of Consulting and Clinical Psychology* 57, no. 3 (1989): 349–57.

CHAPTER 1: THE AMERICAN WAY OF GRIEF

Page

18   *the bereaved were encouraged to scream:* Marilyn Webb, *The Good Death* (New York: Bantam, 1997), p. 295.

19   *1. You have the right to experience your own unique grief:* Alan D. Wolfelt, "The Mourner's Bill of Rights," posted on the Web site of the Center for Loss and Tran-

sition, http://www.centerforloss.com/articles.php?file =mourners.php, accessed 8/20/2010.

20 *As of 2005, Cruse had 5,400 volunteers:* Colin Murray Parkes, "Bereavement Following Disasters," in *Handbook of Bereavement Research and Practice,* edited by Margaret S. Stroebe, Robert O. Hansson, Henk Schut, and Wolfgang Stroebe (Washington, D.C.: American Psychological Association, 2008).

22 *American workers get a notoriously low number:* World Tourism Organization, http://www.infoplease.com/ ipa/A0922052.html, accessed 11/15/09.

22 *"We live in a very peculiar":* Testimony reprinted as Elisabeth Kübler-Ross, "Let's Only Talk About the Present," *The New York Times,* January 15, 1973.

22 *"Just as we've relegated the dying":* Sandra M. Gilbert, *Death's Door* (New York: W. W. Norton, 2006).

23 *"Some sent flowers but did not call for weeks":* Meghan O'Rourke, "Good Grief," *The New Yorker,* January 2010.

23 *If that's the case, then O'Rourke's author's note:* Ibid.

24 *almost every household in the South:* http://www.pbs.org/ civilwar/war/, accessed 12/3/09.

24 *"No monument will ever equal":* Henry Ward Beecher's sermon is from *Our Martyr President, Abraham Lincoln* (New York: Tibbals & Whiting, 1865).

25 *he almost wished that he'd never written it:* Walt Whitman, "O Captain! My Captain!" American Treasures of

the Library of Congress, http://www.loc.gov/exhibits/treasures/trm013.html, accessed 11/18/09.

25 *"By the 1880s, a rigorous and detailed system of rules":* Thomas J. Schlereth, *Victorian America: Transformations in Everyday Life, 1876–1915* (New York: Harper-Collins, 1991), p. 296.

25 *"For one year no formal visiting":* Mary Elizabeth Wilson Sherwood, *Manners and Social Usages* (New York: Harper & Brothers, 1887), p. 192.

26 *Embalming, which had been developed:* LeRoy Bowman, *The American Funeral: A Study in Guilt, Extravagance, and Sublimity* (New York: PublicAffairs, 1959), p. 117.

26 *Relatives often gathered around:* Laurel Hilliker, "Letting Go While Holding On: Postmortem Photography as an Aid in the Grieving Process," *Illness, Crisis and Loss* 14, no. 3 (2006): 245–69.

26 *"The Victorian way of death":* Thomas J. Schlereth, *Victorian America: Transformations in Everyday Life, 1876–1915* (New York: HarperCollins, 1991), p. 293.

27 *One of the most popular:* Ann Douglas, "Heaven Our Home: Consolation Literature in the Northern United States," *American Quarterly* 26, no. 5 (1974): 496–515.

27 *Since women were the keepers:* Peter N. Stearns, *Revolutions in Sorrow: The American Experience of Death in Global Perspective* (Boulder: Paradigm Publishers, 2007), p. 33.

28 *"If one did not mourn well":* Sherwood, *Manners and Social Usages.*

29 *The Victorian period of grief came to a close:* Peter N. Stearns, *American Cool: Constructing a Twentieth Century Emotional Style* (New York: New York University Press, 1994), p. 154.

29 *"Grief is self-pity":* Louise B. Willcox, "Facing Death," *Harper's Bazaar,* January 1911, p. 27, cited in David E. Stannard, "Where All Our Steps Are Tending," in Martha V. Pike and Janice Gray Armstrong, eds., *A Time to Mourn: Expressions of Grief in Nineteenth Century America* (New York: Museums at Stony Brook, 1980).

29 *In England, extended mourning:* Geoffrey Gorer, *Death, Grief, and Mourning* (Garden City, N.Y.: Doubleday, 1965), p. xxii.

29 *"those millions bereaved by the present war":* "The New Mien of Grief," *The Literary Digest,* February 5, 1916.

30 *"Exclusive and important as you may feel":* Toni Torrey, *Wisdom for Widows* (New York: E. P. Dutton, 1941), p. 12.

30 *"The natural processes of corruption and decay":* Geoffrey Gorer, "The Pornography of Death," reprinted in Appendix, Gorer, *Death, Grief, and Mourning.*

31 *"One must avoid":* Philippe Ariès, *Western Attitudes Towards Death* (Baltimore: Johns Hopkins University Press, 1974), p. 87.

31　*"Instead of allowing the awareness of death":* Erich Fromm, *The Sane Society* (New York: Rinehart, 1955), pp. 245–46.

32　*That same year, the first regular course:* That first course was introduced by Robert Fulton at the University of Minnesota according to Lewis R. Aiken, *Dying, Death and Bereavement* (Mahwah, N.J.: Lawrence Erlbaum, 2001).

32　*thanks to Gail Sheehy's* Passages: "Behavior: Passages II," *Time,* August 14, 1978.

33　*"growing league of professionals":* Jane E. Brody, "Dignity for the Dying and Those Around Them Is Goal of New Studies," *The New York Times,* May 3, 1971.

33　*Stand-alone centers for grief:* Georgia Dullea, "A Center for Widows Guides Them Gently over Stages of Grief," *The New York Times,* January 24, 1977.

33　*"common-sense guide for mourning":* Annie Gottlieb, "Widow; by Lynn Caine," *The New York Times,* June 9, 1974.

33　*"Death's now selling books":* Jill Lepore, quoting *Publishers Weekly* in "The Politics of Death," *The New Yorker,* November 30, 2009.

34　*more material on death and dying had appeared:* George E. Dickinson, "Death Education in Medical Schools in the United States," in *Education of the Medical Student in Thanatology,* edited by Bernard Schoenberg (New York: Arno Press, 1981), p. 58.

34 *"It is much too easy to write about death"*: Samuel Vaisrub, "Dying Is Worked to Death," *JAMA* 229, no. 14 (1974): 1909–10.

35 *"We, the living"*: Maya Lin's original proposal for the Vietnam Veterans Memorial, American Treasures of the Library of Congress Web site, http://www.loc.gov/exhibits/treasures/images/tlc0135.jpg, accessed 12/4/09.

36 *Until the Korean war, bodies of U.S. servicemen*: American Battle Monuments Commission Web site, accessed 7/10/10.

36 *Defense Secretary Robert M. Gates reversed the ban*: Elisabeth Bumiller, "Pentagon to Allow Photos of Soldiers' Coffins," *The New York Times*, February 26, 2009.

36 *"While during the Vietnam War grieving"*: Carol Acton, *Grief in Wartime* (New York: Palgrave Macmillan, 2007), p. 176.

38 *"alleviating the responsibility"*: National Funeral Directors Association Guide to Advance Funeral Planning, published by the NFDA, 2007.

38 *"memorial mania"*: Erika Lee Doss, *Memorial Mania: Public Feeling in America* (Chicago: University of Chicago Press, 2010).

39 *In a survey conducted in 1970*: Bert Hayslip Jr. and Cynthia A. Peveto, *Cultural Changes in Attitudes Toward Death, Dying and Bereavement* (New York: Springer, 2005), p. 92.

CHAPTER 2: IS WIDOWHOOD FOREVER?

Page

45   *"The reality is that you will grieve forever"*: Elisabeth Kübler-Ross and David Kessler, *On Grief and Grieving* (New York: Scribner, 2005), p. 230.

46   *"The public wants you to live up to"*: Susan Faludi, *The Terror Dream* (New York: Metropolitan Books, 2007), p. 107.

46   *"After Ed's death, Laura changed"*: Maria Alvarez, Adam Miller, and Andy Geller, "9/11 Widow Remarries," *New York Post,* April 14, 2002.

47   *"If these two weren't having an affair"*: http://www.freerepublic.com/focus/fr/665688/posts#comment, accessed 5/15/09.

47   *Thirty years ago:* Stephen R. Shuchter, *Dimensions of Grief: Adjusting to the Death of a Spouse* (San Francisco: Jossey-Bass, 1986), p. 340.

48   *In 1964, Colin Murray Parkes, a psychiatrist:* C. Murray Parkes, "The Effects of Bereavement on Physical and Mental Health: A Study of the Case Records of Widows," *British Medical Journal* 2 (1964): 274–79.

48   *the most common pattern:* Kathrin Boerner, Camille B. Wortman, and George A. Bonanno, "Resilient or at Risk? A 4-Year Study of Older Adults Who Initially Showed High or Low Distress Following Conjugal Loss," *Journal of Gerontology* 60B, no. 2 (2005): 67–73.

48 *people still continue to think about and miss:* Katherine Carnelley, Camille B. Wortman, Niall Bolger, and Christopher T. Burke, "The Time Course of Grief Reactions to Spousal Loss," *Journal of Personality and Social Psychology* 91, no. 3 (2006): 476–92.

49 *"Grief has no distance":* Joan Didion, *The Year of Magical Thinking* (New York: Vintage, 2006), p. 27.

50 *"In this misguided act of exhibitionism":* John Lahr, "Designated Mourner," *The New Yorker,* April 9, 2007, p. 75.

52 *George Bonanno asked his subjects about the quality of their marriages:* George A. Bonanno, Camille B. Wortman, Darrin R. Lehman, Roger G. Tweed, Michelle Haring, John Sonnega, Deborah Carr, and Randolph M. Nesse, "Resilience to Loss and Chronic Grief: A Prospective Study from Preloss to 18-Months Postloss," *Journal of Personality and Social Psychology* 83, no. 5 (2002): 150–64.

52 *the quick recoverers:* Boerner, Wortman, and Bonanno, "Resilient or at Risk? A 4-Year Study of Older Adults Who Initially Showed High or Low Distress Following Conjugal Loss," 67–73.

52 *Bonanno followed the group:* Ibid.

54 *Resilient grievers appear better equipped to accept death:* Ibid.

54 *early difficulties:* Dale A. Lund, Michael S. Caserta, and Margaret F. Diamond, "The Course of Spousal Be-

reavement in Later Life," in *Handbook of Bereavement: Theory, Research, and Intervention,* edited by Margaret S. Stroebe, Wolfgang Stroebe, and Robert O. Hansson (New York: Cambridge University Press, 1993).

55 *securely attached people were less angry:* Tracey D. Waskowic and Brian M. Chartier, "Attachment and the Experience of Grief Following the Loss of a Spouse," *Omega* 47, no. 1 (2003): 77–91.

55 *researchers compared large groups:* Sara Wilcox, Kelly R. Evenson, Aaron Aragaki, Sylvia Wassertheil-Smoller, Charles P. Mouton, and Barbara Lee Loevinger, "The Effects of Widowhood on Physical and Mental Health, Health Behaviors, and Health Outcomes: The Women's Health Initiative," *Health Psychology* 22, no. 5 (2003): 513–22.

55 *German psychologist Martin Pinquart:* Martin Pinquart, "Loneliness in Married, Widowed, Divorced and Never-Married Older Adults," *Journal of Social and Personal Relationships* 20, no. 1 (2003): 31–53.

56 *The only current study I was able to find:* Karen C. Holden, Jeungkun Kim, and Angela Fontes, "Happiness as a Complex Financial Phenomenon: The Financial and Psychological Adjustment to Widowhood in the U.S." The paper used data from the Wisconsin Longitudinal Study of the University of Wisconsin–Madison and was prepared for a presentation at a conference of the Foundation for International Studies

on Social Security in 2007, but was not published in a peer-reviewed journal.

57 *"very few widows ever seemed"*: Phyllis R. Silverman, *Widow-to-Widow* (New York: Springer, 1986), p. 196.

57 *"Since grief is unending"*: Ibid., p. 7.

57 *"At each stage"*: Ibid., p. 14.

57 *"It may take as long"*: Ibid., p. 201.

58 *"While men need others"*: Ibid., p. viii.

58 they were *"correct in their appraisal"*: Ibid., p. 136.

59 *"The grief process was accepted"*: Julie Ann Wambach, "The Grief Process as a Social Construct," *Omega* 16, no. 3 (1985): 201–11.

59 *"Widows are expected to be devastated"*: Helena Znaniecki Lopata, *Widowhood in an American City* (Cambridge, Mass.: Schenkman, 1973), pp. 75, p. 54.

60 *Half of her respondents agreed with the statement*: Ibid., pp. 75–77.

60 *Why do male widowers get remarried more frequently*: Ken R. Smith, Cathleen D. Zick, and Greg J. Duncan, "Remarriage Patterns Among Recent Widows and Widowers," *Demography* 28, no. 3 (August 1991): 361–74.

61 *"Widowhood was an opportunity for self-expression"*: Robert Di Giulio, *Beyond Widowhood* (New York: Free Press, 1989), pp. 37, 89.

62 *Di Giulio himself remarried*: Death notice for Robert Di Giulio, *The Barre Montpelier Times Argus,* January 28, 2009.

CHAPTER 3: THE WORK OF GRIEF

Page

63 *"In the negotiations"*: LeRoy Bowman, *The American Funeral* (New York: PublicAffairs, 1959), p. 53.

64 *the cremation rate has risen:* "2007 Statistics and Projections to the Year 2025: 2008 Preliminary Data," provided to the author by the Cremation Association of North America in 2009.

67 *"The prescription to tackle grief work"*: Susan J. Zonnebelt-Smeenge and Robert C. DeVries, *Getting to the Other Side of Grief* (Grand Rapids, Mich.: Baker, 1998), p. 65.

67 *"unmanifested grief will be found expressed to the full"*: Helene Deutsch, "Absence of Grief," *The Psychoanalytic Quarterly* 6 (1937): 12–22.

67 *widows who avoided confronting their loss:* Margaret Stroebe and Wolfgang Stroebe, "Does 'Grief Work' Work?," *Journal of Consulting and Clinical Psychology* 59, no. 3 (1991): 479–82. See also R. J. Russac, Nina S. Steighner, and Angela I. Canto, "Grief Work Versus Continuing Bonds: A Call for Paradigm Integration or Replacement?," *Death Studies* 26, no. 6 (2002): 463–78.

68 *talking about the death of a marital partner:* Margaret Stroebe, Wolfgang Stroebe, Henk Schut, Jan van den Bout, and Emmanuelle Zech, "Does Disclosure of Emotions Facilitate Recovery from Bereavement?,"

Journal of Consulting and Clinical Psychology 70, no. 1 (2002): 169–78. See also Wolfgang Stroebe, Henk Schut, and Margaret Stroebe, "Grief Work, Disclosure and Counseling: Do They Help the Bereaved?," *Clinical Psychology Review* 25, no. 4 (June 2005): 395–414.

68 *"[Grief work] demands much more":* Therese A. Rando, *How to Go On Living When Someone You Love Dies* (Lexington, Mass.: Lexington, 1998), p. 16.

69 *"[Denial] is nature's way":* Elisabeth Kübler-Ross and David Kessler, *On Grief and Grieving* (New York: Scribner, 2005), pp. 10, 12, 21.

69 *the six Rs:* Therese A. Rando, *Treatment of Complicated Mourning* (Champaign, Ill.: Research Press, 1993).

71 *there were vast fluctuations:* Toni L. Bisconti, C. S. Bergeman, and Steven M. Boker, "Emotional Well-Being in Recently Bereaved Widows: A Dynamical Systems Approach," *Journal of Gerontology* 59B, no. 4 (2004): 158–67.

72 *In 2008, psychologist Dale Lund:* Dale A. Lund, Rebecca Utz, and Michael S. Caserta, "Humor, Laughter, and Happiness in the Daily Lives of Recently Bereaved Spouses," *Omega* 58, no. 2 (2008): 87–105.

73 *those who were able to do so six months:* George A. Bonanno and Dacher Keltner, "Facial Expressions of Emotion and the Course of Conjugal Bereavement," *Journal of Abnormal Psychology* 106, no. 1 (1997): 126–37.

73   *"It is that respite from the trench of sadness"*: George A. Bonanno, *The Other Side of Sadness* (New York: Basic Books, 2009), p. 43.

74   *"human desire to make sense"*: Holly G. Prigerson and Paul K. Maciejewski, "Grief and Acceptance as Opposite Sides of the Same Coin: Setting a Research Agenda to Study Peaceful Acceptance of Loss," *British Journal of Psychiatry* 193 (December 2008): 453–57.

74   *jumped the track to all sorts of transitional life events:* See Azmy I. Ibrahim, "The Process of Divorce," *Conciliation Courts Review* 22, no. 1 (June 1984): 81–87.

74   *Loyola University in Maryland:* Web site for Loyola University, http://www.loyola.edu/campuslife/healthservices/counselingcenter/hmsick.html, accessed 12/15/09.

75   *"I realized that certain phrases were being expressed"*: Vivienne Cass, "Who Is Influencing Whom? The Relationship Between Identity, Sexual Orientation, and Indigenous Psychologies," *Gay and Lesbian Issues and Psychology Review* 1, no. 2 (2005): 47–52.

76   *"helped them to understand their client"*: Ibid.

76   *The Cass Theory:* Vivienne C. Cass, "Homosexuality Identity Formation: A Theoretical Model," *Journal of Homosexuality* 4, no. 3 (1979): 219–35.

77   *They conducted in-depth interviews with twelve women:* Suzanne Degges-White, Barbara Rice, and Jane E. Myers, "Revisiting Cass' Theory of Sexual Identity Formation: A Study of Lesbian Development," *Journal*

of Mental Health Counseling 22, no. 4 (October 2000): 318–33.

78 *"Theories provide a sense":* Ibid., p. 331.

80 *"Jacqueline tries to grab a cigarette case":* Ruth M. Beard, *An Outline of Piaget's Developmental Psychology for Students and Teachers* (New York: Basic Books, 1969), p. 24.

81 *they are hierarchical:* Ibid., p. 16.

81 *In 1960, Bowlby made waves:* John Bowlby, "Grief and Mourning in Infancy and Early Childhood," *The Psychoanalytic Study of the Child* 15 (1960): 9–52.

82 *the way a child persistently seeks to reunite:* John Bowlby, "Processes of Mourning," *The International Journal of Psychoanalysis* 42 (1961): 317–40.

82 *psychiatrist Colin Murray Parkes:* Inge Bretherton, "The Origins of Attachment Theory: John Bowlby and Mary Ainsworth," *Developmental Psychology* 28, no. 5 (1992): 759–75.

82 *Together, they interviewed twenty-two widows:* John Bowlby and C. Murray Parkes, "Separation and Loss within the Family," in *The Child in His Family* edited by E. James Anthony (New York: Wiley-Interscience, 1970), p. 198.

CHAPTER 4: THE MAKING OF A BESTSELLER

Page

85 *"The voice was very specific":* Virginia Wright, "The Quality of Mercy," *Bates Magazine,* Winter 2002.

85 *"after that lecture, I embarked on a career":* D. Brookes Cowan, "Tribute to Elisabeth Kübler-Ross," published on the Pioneers of Hospice Web site, http://www.pioneersofhospice.org/Tribute%20to%20Elisabeth%20Kubler-Ross.htm, accessed 1/5/10.

86 *one of her former research assistants published:* Dennis Klass and Richard A. Hutch, "Elisabeth Kübler-Ross as a Religious Leader," *Omega* 16, no. 2 (1985–1986): 89–109.

86 *Her nineteenth and final book:* As of June 4, 2010, *On Grief and Grieving* had sold 23,390 hardcover, 41,552 trade paperback, according to Bookscan.

87 *"The fundamental value of this work":* Since the fortieth anniversary edition of *On Death and Dying* (New York: Routledge, 2008) was so expensive (and I already had multiple copies of earlier editions), Allan Kellehear generously e-mailed me his introduction.

87 *"Even though I called it":* Marcia Seligson, "Playboy Interview: Elisabeth Kübler-Ross," *Playboy,* May 1981.

88 *"My experiences have taught me":* Elisabeth Kübler-Ross, *The Wheel of Life* (New York: Touchstone, 1997), p. 16.

89 *"I was weighed, poked, prodded"*: Ibid., p. 29.

90 *"It was the risk one assumed"*: Ibid., p. 142.

90 *"For the next week, I planted myself"*: Ibid., pp. 130–33.

91 *"From an inconspicuous existence"*: Elisabeth Kübler-Ross, Epilogue, in Derek Gill, *Quest: The Life of Elisabeth Kübler-Ross* (New York: Ballantine, 1980), p. 335.

93 *"Dying is still a distasteful"*: Elizabeth Ross [sic], "The Dying Patient as Teacher: An Experiment and an Experience," *Chicago Theological Seminary Register* 57, no. 3 (December 1966): 1–14.

94 *"It is essential to note"*: Allan Kellehear, via e-mail.

95 *"It took three weeks of sitting"*: Kübler-Ross, *The Wheel of Life*, p. 161.

98 *the stages actually did a loop-de-loop*: Parkes also pointed out to me that it was C. Knight Aldrich's 1963 article in *JAMA* that likened a terminally ill person's feelings about his or her own impending death to grief. "I hope you have become aware how much she owed to Professor Aldrich," Parkes e-mailed me. "It was his paper on the dying patient's grief that sparked Elisabeth Kübler-Ross's whole enterprise."

99 *"There was such a lack of chemistry"*: Kübler-Ross, *The Wheel of Life*, p. 176.

99 *Kübler-Ross said that she was ousted*: American Radio Works, "The Hospice Experiment: A Revolution in Dying."

100 *"You cannot stop this work on death"*: Elisabeth Kübler-Ross, *On Life After Death* (Berkeley, Calif.: Celestial Arts, 1991), p. 35.

100 *Nighswonger had replaced Kübler-Ross's five stages:* David Dempsey, "Learning How to Die," *The New York Times Magazine,* November 14, 1971.

100 *he added the stage of "celebration":* John Rea Thomas, *Manual for Naval Hospital Chaplains,* 1970, p. 40; copy provided by the author.

101 *"The important thing, in Nighswonger's opinion":* Dempsey, *The New York Times Magazine.*

101 *he died suddenly of a heart attack:* John Rea Thomas, "A History of Clinical Training and Clinical Pastoral Education in the North Central Region," http://www.ncracpe.org/history&research/ncrhistory.pdf, accessed 1/7/10.

101 *She also began to lecture:* Jane E. Brody, "Dignity in Dying Is Goal of New Studies," *The New York Times,* May 3, 1971.

102 *"By schooling the clinician":* Mark W. Novak and Charles D. Axelrod, "Primitive Myth and Modern Medicine: On Death and Dying," *Psychoanalytic Review* 66 (1979): 443–49.

102 *"Most people, even doctors":* Marjorie C. Meehan, *"On Death and Dying* by Elisabeth Kübler-Ross," *JAMA* 209, no. 5 (August 4, 1969): 776.

103 *"my real job is . . . to tell people":* Kübler-Ross, *On Life After Death,* p. 36.

103 *one of her favorite healers had been pretending:* Kate Coleman, "Afterworld of Entities," *New West,* July 30, 1979.

103 *all of her papers, including twenty-five journals:* Kübler-Ross, *The Wheel of Life,* p. 275.

CHAPTER 5: THE GRIEF COUNSELING INDUSTRY

Page

105 *"I said in the first edition":* J. William Worden, *Grief Counseling and Grief Therapy,* 4th edition (New York: Springer, 2009), p. 9.

106 *As of 2008, approximately five thousand: National Hospice and Palliative Care Organization Facts and Figures,* 2009, http://www.nhpco.org/files/public/Statistics_Re search/NHPCO_facts_and_figures.pdf. Of hospices' paid employees, 4.6 percent are devoted to bereavement, and according to the Centers for Medicare and Medicaid Services, as of April 2009 there were 95,225 full-time employees working at 3,352 Medicare-certified hospices, but this doesn't include nonclinical staff or the employees of approximately 1,500 additional hospices not certified by Medicare; 4.6 percent of 94,225 is 4,830, so 5,000 was a conservative rounding up and doesn't include the 550,000 volunteers. Jon

Radulovic, vice president of communications, National Hospice and Palliative Care Organization, provided me with these numbers.

107 *they surely will at one of the nearly 21,000 funeral homes: 2008 National Directory of Morticians.*

107 *"Funeral homes may use the label":* NFDA *Aftercare Guidelines,* 2010.

108 *anywhere from 20,000 to 100,000 people:* Gordon Thornton of Indiana University of Pennsylvania, the former president of the Association for Death Education and Counseling, who chaired their credentialing committee, estimated 20,000; Robert Neimeyer, professor at the University of Memphis and also a former ADEC president, estimated 100,000.

108 *treated clients are better off than their untreated counterparts:* Michael J. Lambert and Ben M. Ogles, "The Efficacy and Effectiveness of Psychotherapy," in *Bergin and Garfield's Handbook of Psychotherapy and Behavior Change* (New York: Wiley, 2004), pp. 139–93.

108 *they found no consistent pattern of an overall preventive effect:* Joseph M. Currier, Robert A. Neimeyer, and Jeffrey S. Berman, "The Effectiveness of Psychotherapeutic Interventions for Bereaved Persons: A Comprehensive Quantitative Review," *Psychological Bulletin* 134, no. 5 (2008): 648–61.

109 *This is defined as an acute state:* Katherine Shear, Ellen Frank, Patricia R. Houck, and Charles F. Reynolds,

"Treatment of Complicated Grief," *JAMA* 293, no. 21 (June 1, 2005): 2601–3.

110   *Currier and several colleagues:* Joseph M. Currier, Jason M. Holland, and Robert A. Neimeyer, "The Effectiveness of Bereavement Interventions with Children: A Meta-Analytic Review of Controlled Outcome Research," *Journal of Clinical Child and Adolescent Psychology* 36, no. 2 (2007): 253–59.

110   *"It is proposed that this tendency":* Curtis S. Dunkel, "The Association Between Thoughts of Defecation and Thoughts of Death," *Death Studies* 33, no. 4 (2009): 356–71.

113   *"The tasks concept is much more consonant":* Worden, *Grief Counseling and Grief Therapy,* p. 38.

114   *Worden has changed this last task significantly:* Ibid., pp. 39–53.

114   *those with the strongest "continuing bonds":* George A. Bonanno, Eval Gal-Oz, and Nigel P. Field, "Continuing Bonds and Adjustment at 5 Years After the Death of a Spouse," *Journal of Consulting and Clinical Psychology* 71, no. 1 (2003): 110–17.

115   *But as the study of death and dying:* Vanderlyn R. Pine, "A Socio-Historical Portrait of Death Education," *Death Education* 1 (1977): 57–84.

115   *its membership grew dramatically:* Trish Hall, "Solace After Bereavement: Counseling Services Grow," *The New York Times,* Sunday, May 20, 1990. As of Octo-

ber 2009, membership of ADEC was 1,717, according to correspondence with administrative director Emily Burch.

116 *analyzed all the articles on grieving:* Laurel Hilliker, "The Reporting of Grief by One Newspaper of Record for the U.S.: *The New York Times,*" *Omega* 57, no. 3 (2008): 261–78.

116 *"Are our priests and rabbis not up to the task?":* Sally Satel, "An Overabundance of Counseling?," *The New York Times,* April 23, 1999.

121 *"exploded with the stored grief material":* John Shep Jeffreys, *Helping Grieving People When Tears Are Not Enough: A Handbook for Care Providers* (New York: Routledge, 2005), p. 5.

122 *But as John R. Jordan:* John R. Jordan, "Research That Matters: Bridging the Gap Between Research and Practice in Thanatology," *Death Studies* 24, no. 6 (2000): 457–67.

122 *"Mental health professionals in general":* Ibid., p. 461.

123 *So in 2004, he did his own review:* Amanda L. Forte, Malinda Hill, Rachel Pazder, and Chris Feudtner, "Bereavement Care Interventions: A Systematic Review," *BMC Palliative Care* 3, no. 3 (2004).

CHAPTER 6: THE GRIEF DISEASE AND RESILIENCE

Page

127 *Rando asserted that Americans:* Therese A. Rando, "The Increasing Prevalence of Complicated Mourning: The Onslaught Is Just Beginning," *Omega* 26, no. 1 (1992–1993): 43–49. This article was adapted from the keynote address of the same name presented at the thirteenth Annual ADEC conference in 1991.

128 *Grief made its first appearance in the* DSM-III: *Diagnostic and Statistical Manual of Mental Disorders,* 3rd edition (Washington, D.C.: American Psychiatric Association, 1980).

129 *practitioners are given a means to fudge: DSM-IV-TR* (Washington, D.C.: American Psychiatric Association, 2000), pp. 311–12. The guidelines are: "The diagnosis of Major Depressive Disorder is generally not given unless the symptoms are still present 2 months after the loss."

130 *identifying the small percentage of bereaved people:* Holly G. Prigerson and Lauren C. Vanderwerker, "Final Remarks," *Omega* 52, no. 1 (2005–2006): 91–94.

130 *only one treatment has been designed specifically:* Katherine Shear, Ellen Frank, Patricia R. Houck, and Charles F. Reynolds, "Treatment of Complicated Grief: A Randomized, Controlled Trial," *JAMA* 293, no. 21 (2005): 2601–8.

130 *it becomes "pathological" when the survivor:* "Mourning and Melancholia," *The Freud Reader,* edited by Peter Gay (New York: W. W. Norton, 1989), p. 587.

131 *delayed, absent, or excessive grief:* Helene Deutsch, "Absence of Grief," *The Psychoanalytic Quarterly* 6 (1937): 12–22.

131 *picked up on more peculiar deviations:* Erich Lindemann, "Symptomology and Management of Acute Grief," *American Journal of Psychiatry* 101, no. 1 (1944): 141–48.

133 *"A Mother's Grief":* The Daily Mail (London), August 19, 2008.

133 *in the wild, apes often hold on to dead children:* Natalie Angier, "About Death, Just Like Us or Pretty Much Unaware?," *The New York Times,* September 1, 2008.

133 *In the aftermath of his death:* Kenneth J. Doka, "Fulfillment as Sanders' Sixth Phase of Bereavement: The Unfinished Work of Catherine Sanders," *Omega* 52, no. 2 (2005–2006): 143–51.

134 *parents who had lost children:* Catherine M. Sanders, "A Comparison of Adult Bereavement in the Death of a Spouse, Child and Parent," *Omega* 10, no. 4 (1979–1980): 303–20.

135 *the death of a spouse caused longer-lasting depression:* Fran H. Norris and Stanley A. Murrell, "Social Support, Life Events, and Stress as Modifiers of Adjustment to Bereavement by Older Adults," *Psychology and Aging* 5, no. 3 (September 1990): 429–36.

135   *the death of an adult son or daughter:* Marc Cleiren, René
Diekstra, J.F.M. Kerkhof, Jan van der Wal, "Mode of
Death and Kinship in Bereavement: Focusing on 'Who'
Rather than 'How,' " *Crisis* 15, no. 1 (1994): 22–36.

135   *one third of all major losses:* Catherine Sanders, "Risk
Factors in Bereavement Outcome," *Journal of Social Issues* 44, no. 3 (1988): 97–111.

136   *the estimated prevalence figure:* Holly Prigerson and
Selby Jacobs estimated 20 percent prevalence in "Traumatic Grief as a Distinct Disorder," in *Handbook of
Bereavement Research: Consequences, Coping, and Care*
(Washington, D.C.: American Psychological Association, 2001). Holly Prigerson told the author in 2009
that about 10 percent of cases of widowhood from
natural causes result in complicated grief, and George
Bonanno's studies, in George A. Bonanno, Camille B.
Wortman, Darrin R. Lehman, Roger G. Tweed, Michelle Haring, John Sonnega, Deborah Carr, and Randolph M. Nesse, "Resilience to Loss and Chronic Grief:
A Prospective Study from Preloss to 18-Months Postloss," *Journal of Personality and Social Psychology* 83,
no. 5 (2002); and Kathrin Boerner, Camille B. Wortman, and George Bonanno, "Resilient or at Risk? A
4-Year Study of Older Adults Who Initially Showed
High or Low Distress Following Conjugal Loss," *Journal of Gerontology* 60B, no. 2 (2005), indicate a range
from 10 to 15 percent.

136 *The resulting definition:* Holly Prigerson et al., "Consensus Criteria for Traumatic Grief: A Preliminary Empirical Test," *British Journal of Psychiatry* 174 (1999): 67–73.

136 *Even the most sudden, violent incident:* Katherine M. Shear, Carlos T. Jackson, Susan M. Essock, Sheila D. Donahue, and Chip J. Felton, "Screening for Complicated Grief Among Project Liberty Service Recipients 18 Months After September 11, 2001," *Psychiatric Services* 57, no. 9 (2006): 1291–97.

137 *it was because most of New York City remained structurally intact:* Arieh Y. Shalev, "Lessons Learned from 9/11: the Boundaries of a Mental Health Approach to Mass Casualty Events," in *9/11: Mental Health in the Wake of Terrorist Attacks,* edited by Yuval Neria, Raz Gross, Randall Marshall, and Ezra Susser (New York: Cambridge University Press, 2006).

137 *One estimate put the number of therapists:* Christina Hoff Somers and Sally Satel, *One Nation Under Therapy: How the Helping Culture Is Eroding Self-Reliance* (New York: St. Martin's, 2005), p. 178.

137 *these disaster mental health (DMH) volunteers:* Erica Lowry and Gerald McCleery, "The American Red Cross and September 11th Fund Mental Health Disaster Response," in *On the Ground After September 11,* edited by Yael Danieli and Robert L. Dingman (Binghamton, N.Y.: Haworth, 2005).

137 *they were among the few civilians:* Karen M. Seeley, *Therapy After Terror: 9/11, Psychotherapists, and Mental Health* (New York: Cambridge University Press, 2008), p. 2.

137 *They conducted weeks of door-to-door outreach:* Lowry and McCleery, "The American Red Cross and September 11th Fund Mental Health Disaster Response," p. 187.

138 *"While we organized to try and find our loved ones":* Alissa Torres, Sungyoon Choi, *American Widow* (New York: Villard, 2008), p. 91.

139 *"campaigned strongly to have the VIP":* Shiya Ribowsky, "Challenges in Identification: The World Trade Center Dead," in *On the Ground After September 11,* edited by Danieli and Dingman, pp. 80–81.

140 *"I continue to have some skepticism":* Simon Wessely, "What Mental Health Professionals Should and Should Not Do," in *9/11 Mental Health in the Wake of Terrorist Attacks,* edited by Neria, Gross, Marshall, and Susser.

140 *"were advised to discuss their experiences immediately":* Seeley, *Therapy After Terror,* p. 3.

140 *critical incident stress debriefing:* For an in-depth look at the problems of critical incident stress debriefing, see Jerome Groopman, "The Grief Industry," *The New Yorker,* January 26, 2004.

141 *"I have learned about the whacked-out phenomenon":* Danielle Gardner, "A Learning Curve? A Family

Member's Guidebook to Private Grief in Public Tragedy," in *On the Ground After September 11,* edited by Danieli and Dingman, p. 627.

141 *"Those who specialized in bereavement":* Seeley, *Therapy After Terror,* p. 16.

142 *"The mayor was concerned about his messaging":* Neal L. Cohen, "Reflections on the Public Health and Mental Health Response to 9/11," in *On the Ground After September 11,* edited by Danieli and Dingman, p. 26.

142 *people were not so much concerned with ongoing terrorist attacks:* Michael Cohen, "Strategic Communications and Mental Health: The WTC Attacks, 1993 and 2001," in *On the Ground After September 11,* edited by Danieli and Dingman, p. 130.

142 *"The fundamental stages of recovery":* Judith Lewis Herman, *Trauma and Recovery* (New York: Basic Books, 1992), p. 3.

143 *Early the next morning:* Wayne Barrett and Dan Collins, *Grand Illusion* (New York: HarperCollins, 2006), p. 21.

144 *"I believe that it is certainly time to say":* Michael Cooper, "A Nation Challenged: The Trade Center," *The New York Times,* September 25, 2001.

145 *when he used a press conference:* Elisabeth Bumiller, "Giuliani and His Wife of 16 Years Are Separating," *The New York Times,* May 11, 2000.

145 *he also took steps to prevent them:* Cooper, "A Nation Challenged: The Trade Center."

145   *Giuliani announced that he would be giving families:* Amy Waldman, "A Nation Challenged: Mementos," *The New York Times,* October 15, 2001.

145   *"Giving survivors space":* Steven M. Crimando and Gladys Padro, "Across the River: New Jersey's Response to 9/11," in *On the Ground After September 11,* edited by Danieli and Dingman.

146   *They found that out of a sample of nearly one thousand people:* Sandro Galea, Jennifer Ahern, Heidi Resnick, Dean Kilpatrick, Michael Bucuvalas, Joel Gold, and David Vlahov, "Psychological Sequelae of the September 11 Terrorist Attacks in New York City," *New England Journal of Medicine* 346, no. 13 (2002): 982–87.

146   *launched Project Liberty:* Sheila A. Donahue, Carol B. Lanzara, Chip J. Felton, Susan M. Essock, Sharon Carpinello, "Project Liberty: New York's Crisis Counseling Program Created in the Aftermath of September 11, 2001," *Psychiatric Services* 57, no. 9 (2006): 1253–58.

147   *"New York Needs Us Strong":* Neal L. Cohen, "Reflections on the Public Health and Mental Health Response to 9/11," in *On the Ground After September 11,* edited by Danieli and Dingman, p. 27.

147   *People who lost family members accounted for 40 percent:* Nancy H. Covell, Sheila A. Donahue, George Allen, M. Jameson Foster, Chip J. Felton, and Susan M. Essock, "Use of Project Liberty Counseling Services over

Time by Individuals in Various Risk Categories," *Psychiatric Services* 57, no. 9 (2006): 1268–70.

148   *only 9,204 people had enrolled:* Lowry and McCleery, "The American Red Cross and September 11th Fund Mental Health Disaster Response," in *On the Ground After September 11,* edited by Danieli and Dingman, p. 187.

148   *The treatment consisted of the following:* Katherine M. Shear, Ellen Frank, Edna Foa, Christine Cherry, Charles F. Reynolds, III, Joni Vander Bilt, and Sophia Masters, "Traumatic Grief Treatment: A Pilot Study," *The American Journal of Psychiatry* 158, no. 9 (2001): 1506–8.

149   *In the spring of 2002, Shear surveyed:* Katherine M. Shear, Carlos T. Jackson, Susan M. Essock, Sheila A. Donahue, and Chip J. Felton, "Screening for Complicated Grief Among Project Liberty Service Recipients 18 Months After September 11, 2001," *Psychiatric Services* 57, no. 9 (2006): 1291–97.

150   *"It is important to recognize":* Katherine M. Shear, "Project Liberty Enhanced Services Intervention, Traumatic Grief: Guidebook for Enhanced Services Providers," October 2003; provided to the author by April Naturale.

151   *was best defined by the severity and persistence of the reaction:* Jason M. Holland, Robert A. Neimeyer, Paul A. Boelen, and Holly G. Prigerson, "The Underlying

Structure of Grief: A Taxometric Investigation of Prolonged and Normal Reactions to Loss," *Journal of Psychopathology and Behavioral Assessment* 31, no. 3 (2009): 190–201.

151 *Prolonged Grief Disorder (PGD):* Holly G. Prigerson et al., "Prolonged Grief Disorder: Psychometric Validation of Criteria Proposed for DSM-V and ICD-11," *PLoS Medicine* 6, no. 8 (2009).

153 *In 2008, Mary Frances O'Connor:* Mary Frances O'Connor, David K. Wellisch, Annette L. Stanton, Naomi I. Eisenberger, Michael R. Irwin, and Matthew D. Lieberman, "Craving Love? Enduring Grief Activates Brain's Reward Center," *Neuroimage* 42, no. 2 (2008): 969–72.

153 *One predictor is the nature of the survivor's relationship:* Louis A. Gamino, Kenneth W. Sewall, and Larry W. Easterling, "Scott & White Grief Study: An Empirical Test of Predictors of Intensified Mourning," *Death Studies* 22, no. 4 (1998): 333–55.

154 *Certain traits and attitudes may also predispose people:* Ibid.

155 *Zisook and others have run small trials:* Tracey Auster, Christine Moutier, Nicole Lanouette, and Sidney Zisook, "Bereavement and Depression: Implications for Diagnosis and Treatment," *Psychiatric Annals* 38, no. 10 (2008).

155 *As Bonanno defined it in 2004:* George A. Bonanno, "Loss, Trauma, and Human Resilience: Have We Un-

derestimated the Human Capacity to Thrive After Extremely Aversive Events?," *American Psychologist* 59, no. 1 (January 2004): 20–28.

157   *Then Bonanno came across an article:* Camille B. Wortman and Roxane Cohen Silver, "The Myths of Coping with Loss," *Journal of Consulting and Clinical Psychology* 57, no. 3 (1989): 349–57.

158   *The phenomenon was found to be so common:* Ann S. Masten, "Ordinary Magic: Resilience Processes in Development," *American Psychologist* 56 (2001): 227–38.

158   *laughing and smiling are more helpful:* Karin G. Coifman, George A. Bonanno, Rebecca D. Ray, and James J. Gross, "Does Repressive Coping Promote Resilience?," *Journal of Personality and Social Psychology* 92, no. 4 (2007): 745–58. See also Dacher Keltner and George A. Bonanno, "Facial Expressions of Emotion and the Course of Conjugal Bereavement," *Journal of Abnormal Psychology* 106, no. 1 (1997): 126–37.

160   *those who felt the lowest amount of stress:* Nicole E. Rossi, Toni L. Bisconti, and C. S. Bergman, "The Role of Dispositional Resilience in Regaining Life Satisfaction After the Loss of a Spouse," *Death Studies* 31, no. 10 (2007): 863–83.

160   *"Well, I have never let anything":* Ibid., p. 866.

161   *So the military called Martin Seligman:* Benedict Carey, "Mental Stress Training Is Planned for U.S. Soldiers," *The New York Times,* August 17, 2009.

CHAPTER 7: GRIEF AND THE SEXES

Page

163 *"enabled not only women":* Tony Walter, "The New Public Mourning," in *Handbook of Bereavement Research and Practice,* edited by Margaret S. Stroebe, Robert O. Hansson, Henk Schut, and Wolfgang Stroebe (Washington, D.C.: American Psychological Association, 2008).

164 *"The general failure of the literature on death to discuss gender":* Monica McGoldrick and Froma Walsh, editors, *Living Beyond Loss: Death in the Family*, 2nd edition (New York: W. W. Norton, 2004), p. 99.

164 *in which she argued, using slim evidence:* For an eye-opening analysis of the impact of Gilligan's theories and some of the problems with her research, see Rosalind Barnett and Caryl Rivers, *Same Difference: How Gender Myths Are Hurting Our Relationships, Our Children, and Our Jobs* (New York: Basic Books, 2004), pp. 20–45.

166 *two female social workers:* Eunice Gorman and Laura Lewis, "Bereavement Support Group? No Thanks, I'm Dating," presented at the annual conference of the Association for Death Education and Counseling, Dallas, Texas, April 16, 2009.

166 *researchers have focused more on women in general:* For example, Peter Marris, *Widows and Their Families* (New York: Routledge, 1958); and C. Murray Parkes, "Effects of Bereavement on Physical and Mental

Health—A Study of the Medical Records of Widows," *British Medical Journal* 2 (1964): 274–79.

167 *they remarry less frequently:* Ken R. Smith, Cathleen D. Zick, and Greg J. Duncan, "Remarriage Patterns Among Recent Widows and Widowers," *Demography* 28, no. 3 (1991): 361–74.

167 *there were 13.3 million widows:* "Marital Status of the Population Aged 15 and Older by Sex, Age, Race and Hispanic Origin: 1950 to Present," Fertility and Family Statistics Branch, U.S. Census Bureau.

167 *men who were less depressed:* Margaret S. Stroebe and Wolfgang Stroebe, "Who Participates in Bereavement Research? A Review and Empirical Study," *Omega* 20, no. 1 (1989): 1–29.

168 *relatively speaking, men suffer more from being bereaved:* Margaret S. Stroebe, Wolfgang Stroebe, and Henk Schut, "Gender Differences in Adjustment to Bereavement: An Empirical and Theoretical Review," *Review of General Psychology* 5, no. 1 (2001): 62–83.

169 *it was the women who were more likely to declare:* Eugene E. McDowell, Judy M. Stillion, Kenneth J. Doka, Terry L. Martin, and Beth D. Stillion, "Masculine Grief and Feminine Grief: Vive la Différence ou le Manque de Différence," presented at the annual conference of Association for Death Education and Counseling, Chicago, Illinois, March 21, 1998. "Objective and Subjective Responses of 50 Death and Dying Professionals to

Seven Questions Concerning Gender Differences in Grief" tables provided by McDowell to the author.

169 *Women are also more likely to become grief counselors:* Although there is no official tally, an analysis of the membership directory of the Association for Death Education and Counseling shows that the vast majority of members are women.

177 *30 percent of the effect sizes were close to zero:* Janet Shibley Hyde, "The Gender Similarities Hypothesis," *American Psychologist* 60, no. 6 (September 2005): 581–92.

178 *"For many years, we have relied":* Dale A. Lund, editor, *Men Coping with Grief* (Amityville, N.Y.: Baywood, 2000), p. 2.

178 *"This finding does not mean":* Dale A. Lund, Michael S. Caserta, and Margaret F. Dimond, "Gender Differences Through Two Years of Bereavement Among the Elderly," *The Gerontologist* 26, no. 3 (1986): 314–20. See also Michael S. Caserta and Dale A. Lund, "Bereaved Older Adults Who Seek Early Professional Help," *Death Studies* 16, no. 1 (1992): 17–30.

179 *When participants were told:* Steven J. Spencer, Claude M. Steele, and Diane M. Quinn, "Stereotype Threat and Women's Math Performance," *Journal of Experimental Social Psychology* 35, no. 1 (1999): 4–28.

179 *men showed more aggressive tactics:* Jenifer R. Lightdale and Deborah A. Prentice, "Rethinking Sex Differences

in Aggression: Aggressive Behavior in the Absence of Social Roles," *Personality and Social Psychology Bulletin* 20, no. 1 (1994): 34–44.

180 *bereaved fathers, when provided with the veil of anonymity:* George W. Musambira, Sally O. Hastings, and Judith D. Hoover, "Bereavement, Gender, and Cyberspace: A Content Analysis of Parents' Memorials to Their Children," *Omega* 54, no. 4 (2006–2007): 263–79.

CHAPTER 8: GRIEF FOR EXPORT

Page

181 *"Americans are the largest producers":* Stanley Sue, "Science, Ethnicity and Bias: Where Have We Gone Wrong?" *American Psychologist* 54, no. 12 (1999): 1070–77.

181 *"We are engaged in the grand project":* Ethan Watters, *Crazy Like Us: The Globalization of the American Psyche* (New York: Free Press, 2010), p. 1.

182 *The exportation began: Death and Bereavement in Europe,* edited by John D. Morgan and Pittu Laungani (Amityville, N.Y.: Baywood, 2004).

182 *her former employees continue to teach:* http://www.externalizationworkshops.com/, accessed 4/21/10.

182 *"it has been an article of faith":* Allan Kellehear, "The Australian Way of Death," in *Death and Bereavement*

in Asia, Australia and New Zealand, edited by John D. Morgan and Pittu Laungani (Amityville, N.Y.: Baywood, 2005), p. 18.

182 *"The speaker focuses on the loss experience":* Alfons Deeken, "Grief Education and Bereavement Support in Japan," *Psychiatry and Clinical Neurosciences* 49, no. S1 (March 17, 2008): 129–33.

183 *there are now two centers:* Those centers are the Comfort, Care, Concern Group and the Jessie and Thomas Tam Center of the Society for the Promotion of Hospice Care.

183 *One of the first grief support groups in Hong Kong:* Cecilia L. W. Chan and Amy Y. M. Chow, "An Indigenous Psycho-Educational Group for Chinese Bereaved Family Members," *Hong Kong Journal of Social Work* 32, no. 1 (1998): 1–20.

183 *surveyed seventy-eight different ethnic groups:* Paul C. Rosenblatt, *Grief and Mourning in Cross-Cultural Perspective* (New Haven: Human Relations Area Files Press, 1976).

184 *a years-long study of bereavement in Egypt and Bali:* Unni Wikan, "Bereavement and Loss in Two Muslim Communities: Egypt and Bali Compared," *Social Science and Medicine* 27, no. 5 (1988): 451–60.

186 *"We'd visited the village without warning":* David Brooks, "Where's the Trauma and the Grief?," *The New York Times,* August 14, 2008.

187 *"I even asked him to go"*: Amy Y. M. Chow, Cecilia L. W. Chan, Samuel M. Y. Ho, Doris M. W. Tse, Margaret H. P. Suen, and Karen F. K. Yuen, "Qualitative Study of Chinese Widows in Hong Kong: Insights for Psycho-Social Care in Hospice Settings," *Palliative Medicine* 20, no. 5 (2006): 513–20.

187 *As of 2008, there were almost two million people:* Migration Policy Institute Web site, accessed 5/10.

187 *"It is not that Asian-Americans were unwilling":* Stanley Sue, Nathaniel Wagner, Davis Ja, Charlene Margullis, and Louise Lew, "Conceptions of Mental Illness Among Asian and Caucasian-American Students," *Psychological Reports* 38 (1976): 703–8.

188 *these immigrants had held on to coping strategies:* Christine J. Yeh, Arpana C. Inman, Angela B. Kim, and Yuki Okubo, "Asian-American Families' Collectivistic Coping Strategies in Response to 9/11," *Cultural Diversity and Ethnic Minority Psychology* 12, no. 1 (January 2006): 134–48.

194 *"Hey, noble one!":* Robert A. F. Thurman, translator, *The Tibetan Book of the Dead,* composed by Padma Sambhava (New York: Bantam, 1994).

195 *"Not from weeping":* From the Parable of the Mustard Seed, *The Gospel of Buddha,* compiled by Paul Carus (Chicago: Open Court, 1894; reprint, 2004).

AFTERWORD

Page

197   *"No culture before has abandoned":* Tony Walter, *On Be-reavement* (Philadelphia: Open University Press, 1999), p. 166.

Index

Index

attachment theory, 81
Authentic Happiness (Seligman),
  161–62

baby boomers, 32–33, 98
Bali, 184–85
Barham, Jay, 87
Beamer, Lisa, 46
Beattie, Melody, 21
Beecher, Henry Ward, 24
Benkong, Shih, 192–95
Beyond Widowhood (Di Giulio),
  61
Bisconti, Toni, 71–72, 73–74,
  160–61
Blair, Pamela D., 21
Bonanno, George, 13, 15, 48,
  49, 51–52, 73, 114, 155–60,
  165
Bonanno, Paulette, 157
Bournakel, Nicos, 52–53
Bournakel, Sarah White, 52–53,
  198–200
Bournakel, Stefan, 52–53, 199
Bowlby, John, 81–82, 97, 98
Bowman, LeRoy, 63–64
Braestrup, Kate, 7
Breitwieser, Kristin, 46
Bridge Program, 190–91
British Journal of Psychiatry, 10,
  74, 136
Brody, Jane, 33
Brooks, David, 186
Bruce, Lenny, 83

Buddhism, 186, 192–95
Bush, George H. W., 36

Caine, Lynn, 33
Canfield, Jack, 21
Cass, Vivienne, 74–79
Census Bureau, U.S., 167
Center for Attitudinal Healing,
  53
Center for Death Education at the
  University of Michigan, 34
Center for Grief Recovery, 107
Center for Loss and Life Transi-
  tion, 18
Center for the Advancement of
  Health, 12
Center for Urban Epidemiologi-
  cal Studies, 146
Charles B. Wang Community
  Health Center, 190
Chen, Teddy, 190–91
Chicago Theological Seminary,
  92–93, 95
children, death of, 132–35
China, 186–87, 188
Citizen's Advice Bureau, 20
Civil War, U.S., 24, 26, 27
Claudio (gorilla), 133
Clinton, Hillary, 1
Cobb, A. Beatrix, 96–97
Codependent No More (Beattie), 21
cognitive behavioral counseling,
  109
Cohen, Michael, 142–45

Index

Index

Index